Robert Zahner

The Transmission of Power by Compressed Air

Robert Zahner

The Transmission of Power by Compressed Air

ISBN/EAN: 9783337814243

Printed in Europe, USA, Canada, Australia, Japan

Cover: Foto ©berggeist007 / pixelio.de

More available books at **www.hansebooks.com**

THE
TRANSMISSION OF POWER

BY

COMPRESSED AIR.

BY

ROBERT ZAHNER, M.E.

REPRINTED FROM VAN NOSTRAND'S MAGAZINE.

NEW YORK:
D. VAN NOSTRAND, PUBLISHER,
23 MURRAY AND 27 WARREN STREET.
1878.

PREFACE.

THE subject of Compressed Air and Compressed Air Machinery offers a wide field for useful investigation. It has been attempted in these pages to add something to the scanty stock of its literature.

Compressed Air has become a most efficient and powerful agent in the hands of the modern engineer. Its applicacations are rapidly growing, both in extent and importance. The subject demands careful attention and study.

There can be no doubt that the great waste of energy that to-day accompanies the use of Compressed Air is due, not only to sickly design and faulty ·construction of machines, but very largely also to the general ignorance of the principles of thermodynamics. Hence, we have started with the general equation of thermodynamics, which expresses the relation between heat and mechanical energy under all circumstances, and have deduced from it all the formulas and results necessary to an intelligent comprehension of our subject. We have not tried to be simple by avoiding the higher mathematical analysis; but *thorough* and *clear*, by beginning at the very bottom and fully explaining every step and principle. It will not be necessary for the average reader to study any work on thermodynamics or higher mathematics, preparatory to reading this.

Zeuner's "Théorie Méchanique de la Chaleur," Clausius, Rankine and McCullough on Heat, Riedler's "Luftcompressions-maschinen," have been freely used. The works of MM. Cornet, Mallard and Pochet and others, have all, I believe, received credit in the text.

R. Z.

STEVENS INST. OF TECHNOLOGY,
June, 1878.

TABLE OF CONTENTS.

PREFACE.

INTRODUCTION.

I, Historical Notice; II, Its Applications and Future.

CHAPTER I.—THE CONDITIONS MODIFYING EFFICIENCY IN THE USE OF COMPRESSED AIR.

1. Loss of Energy.
2. Methods of Cooling.
3. Conditions Most Favorable to the Highest Efficiency.
4. Efficiency Attained in Practice.
5. Efficiency of Full Pressure and Complete Expansion Compared.
6. Losses of Transmission.

CHAPTER II.—PHYSICAL PROPERTIES AND LAWS OF AIR.

1. Introductory
2. Boyle's Law.

3. The Law of Gay-Lussac.
4. Boyle's and Gay-Lussac's Law.
5. Absolute Temperature.
6. The Law of Pressure, Density and Temperature.
7. The Measurement of Heat.

CHAPTER III.—THERMODYNAMIC PRINCIPLES AND FORMULAS.

1. Introductory.
2. Heat and Temperature.
3. The Two Laws of Thermodynamics.
4. Heat and Mechanical Energy.
5. The Differential Equation of the Second Law.

CHAPTER IV.—THERMODYNAMIC EQUATIONS APPLIED TO PERMANENT GASES.

1. The Determination of the Specific Heat at Constant Volume.
2. Internal Heat.
3. Quantity of Heat Supplied.
4. Expansion with Temperature Constant.
5. Expansion in a Perfectly Non-Conducting Cylinder.
6. Variations in the Temperature of a Gas during Adiabatic Compression or Expansion.

CHAPTER V.—THERMODYNAMIC LAWS APPLIED TO THE ACTION OF COMPRESSED AIR.

1. Fundamental Formulas.
2. Work Spent in Compression.
3. Work Obtainable from Compressed Air.
4. The Theory of Compression.
5. The Theory of Transmission.
6. The Theory of Complete Expansive Working.
7. The Theory of Full Pressure Working.
8. The Theory of Incomplete Expansive Working.
9. Graphical Representation for the Action of Compressed Air.

CHAPTER VI.—THE EFFICIENCY THEORETICALLY ATTAINABLE.

1. The Efficiency of the Compressor and the Compressed Air Engine as a System.
2. Maximum Efficiency calculated from the Indicated Work.
3. Efficiency of Full Pressure and Complete Expansion Compared.

CHAPTER VII.—THE EFFECTS OF MOISTURE, OF THE INJECTION OF WATER AND OF THE CONDUCTION OF HEAT.

1. General Statement.
2. Effects of Moisture.
3. The Injection of Water.

CHAPTER VIII.—AMERICAN AND EUROPEAN AIR COMPRESSORS.
1. Pump Compressors.
2. Single Acting Wet Compressors.
3. Double and Direct Acting Compressors.
4. Design and Construction.

CHAPTER IX.—EXAMPLES FROM PRACTICE.
1. Republic Iron Company.
2. Economy Promoted by the use of Compressed Air.
3. Compressed Air Motor Street Cars.

TABLES. Page.

I. Values of $\frac{\tau}{\tau_0}$, $\frac{v_0}{v}$, &c., for convenient values of $\frac{p}{p_0}$ 72, 73
II. Final Temperatures of Compressed Air............................. 78
III. Final Temperature of after Expansion 84
IV. Theoretical Efficiencies............ 97
V. Theoretical Efficiencies for Full Pressure and Complete Expansion. 100
VI. Effects of Moisture on final Temperature 105
VII. Work of Isothermal and Adiabatic-Compression.................... 108
VIII. Quantity of Water necessary in Compression.................... 110
IX. Quantity of Water to be Injected into the Working Cylinder....... 112

TRANSMISSION OF POWER BY COMPRESSED AIR.

INTRODUCTION.

I.

HISTORICAL NOTICE.

The application of compressed air to industrial purposes dates from the close of the last century. Long before this, indeed, we find isolated attempts made to apply it in a variety of ways; but its final success must be ascribed to the present age—the age of mechanic arts—an age inaugurated in so splendid a manner by the genius of Watt, and which has been so wonderfully productive in good to mankind.

Without going into any details as to its history, we shall only name the English engineers, Cubitt and Brunell, who, in 1851–4, first applied compressed air in its statical application to the sinking of bridge caissons; the Genoese Professor, M. Collodon, who, in 1852, first conceived and suggested the idea of employing it in the proposed tunneling of the Alps; and, finally, the distinguished French engineer, Sommeiller, who first practically realized and applied Collodon's idea in the boring of the Mt. Cenis Tunnel.

II.

ITS APPLICATIONS AND ITS FUTURE.

The applications of compressed air are very numerous, its most important one being the transmission of power by its means.

Custom has confined the term "transmission of power" to such devices as are employed to convey power from one place to another, without including organized machines through which it is directly applied to the performance of work.

Power is transmitted by means of shafts, belts, friction-wheels, gearing, wire-rope, and by water, steam and air. There is nothing of equal importance connected with mechanical engineering in regard to which there exists a greater diversity of opinion, or in which there is a greater diversity of practice, than in the means of transmitting power. Yet in every case it may be assumed that some particular plan is better than any other, and that plan can be best determined by studying, first, the principles of the different modes of transmission and their adaptation to the special conditions that exist; and, secondly, precedents and examples.

For transmitting power to great distances, shafts, belts, friction-wheels and gearing are clearly out of the question. The practical incompressibility and want of elasticity of water, renders the hydraulic method unfit for transmitting regularly a constant amount of power; it can be used to advantage only where motive power, acting continuously, is to be accumulated and applied at intervals, as for raising weights, operating punches, compressive forging and other work of an intermittent character, requiring a great force acting through a small distance.

Whether steam, air or wire-rope is to be made the means of transmitting power from the prime-mover to the machine, depends entirely upon the special conditions of each case. In carrying steam to great distances very important losses occur from condensation in the pipes; especially during cold weather. The wear and tear of cables lessen the advantages of the telodynamic transmission; steep inclinations and frequent changes of direction of the line of transmission often exclude its adoption; while it is entirely excluded when it is rather a question of distributing a small force over a large number of points than of concentrating a large force at one or two points.

Compressed air is the only general mode of transmitting power; the only one that is always and in every case possible, no matter how great the distance nor how the power is to be distributed and applied. No doubt as a means of utilizing distant, yet hitherto unavailable, sources of power, the importance of this medium can hardly be overestimated.

But compressed air is also a *storer* of power, for we can accumulate any desired pressure in a reservoir situated at any distance from the source, and draw upon this store of energy at any time; which is not possible either in the case of steam, water or wire-rope.

Larger supply-pipes are required for steam or water transmission; the inconveniences resulting from hot steam pipes, the leakages in water pipes, the high velocities required in telodynamic transmission are all without their counterparts in compressed-air transmission. Compressed air is furthermore independent of differences of level between the source of power and its points of application, and is perfectly applicable no matter how winding and broken the path of transmission.

But especially is compressed air adapted to underground work. Steam is here entirely excluded; for the confined character of the situation and the difficulty of providing an adequate ventilation, render its use impossible; compressed air, besides being free from the objectionable features of steam, possesses properties that render its employment conducive to coolness and purity in the atmosphere into which it is exhausted. The boring of such tunnels as the Mt. Cenis and St. Gothard would have been impossible without it. Its easy conveyance to any point of the underground workings; its ready application at any point; the improvement it produces in the ventilating currents; the complete absence of heat in the conducting pipes; the ease with which it is distributed when it is necessary to employ many machines whose positions are daily changing, such as hauling engines, coal-cutting machines and portable rock-drills; these, and many other advantages, when contrasted with steam under like conditions, give compressed air a value which the engineer will fully appreciate.

There is every reason to believe that

compressed air is to receive a still more extensive application. The diminished cost of motive power when generated on a large scale, when compared with that of a number of separate steam engines and boilers distributed over manufacturing districts, and the expense and danger of maintaining an independent steam power for each separate establishment where power is used, are strong reasons for generating and distributing compressed air through mains and pipes laid below the surface of streets in the same way as gas and water are now supplied. Especially in large cities would the benefits of such a system be invaluable; no more disastrous boiler explosions in shops filled with hundreds of working men and women; the danger of fire greatly reduced; a corresponding reduction in insurance rates; an important saving of space; cleanliness, convenience and economy. We say economy! For there is no doubt that a permanently located air-compressing plant, established on a large scale, and designed on principles of true economy and not with reference to cheapness of construction, would supply power at a much less cost than is supposed. Besides, there are many natural sources of power, as water power, which could by this means be utilized, and their immense stores of energy conveyed to the great centers of business and manufacture.

As affording a means of dispensing with animal power on our street railroads, compressed air has been proposed as the motor to drive our street cars. It has already met with some success in this direction, and, to-day, there are eminent French, English and American engineers at work upon this interesting problem.

The compressed-air locomotives of M. Ribourt, now in use at the St. Gothard Tunnel, give very satisfactory results. They are compact, neat and comparatively economical.

Compressed air is also applied in a variety of other ways; in signaling, in propelling torpedo boats; in ventilating large and confined spaces; in driving machinery in confined shops; in sinking bridge caissons. The pneumatic dispatch system, the air brake, the pneumatic elevator and hoist, are further examples of its use.

CHAPTER I.

THE CONDITIONS MODIFYING EFFICIENCY IN THE USE OF COMPRESSED AIR.

I.

LOSS OF ENERGY.

What is at present required in the use of compressed air is a considerable diminution in the first cost of obtaining it by really improving the compressor, and a practical means of working it at a high rate of expansion without the present attendant losses. In the best machines in use at the present day, the *useful effect*, that is, the ratio of the work done by the air to that done upon it, is very small. The losses are chiefly due to the following causes:

1. The compression of air develops heat; and as the compressed air always

cools down to the temperature of the surrounding atmosphere before it is used, the mechanical equivalent of this dissipated heat is work lost.

2. The heat of compression increases the volume of the air, and hence it is necessary to carry the air to a higher pressure in the compressor in order that we may finally have a given volume of air at a given pressure, and at the temperature of the surrounding atmosphere. The work spent in effecting this excess of pressure is work lost.

3. The great cold which results when air expands against a resistance, forbids expansive working, which is equivalent to saying, forbids the realization of a high degree of efficiency in the use of compressed air.

4. Friction of the air in the pipes, leakage, dead spaces, the resistance offered by the valves, insufficiency of valve-area, inferior workmanship and slovenly attendance, are all more or less serious causes of loss of power.

The question now is, how can we get

rid of these losses and obtain a higher efficiency?

The first cause of loss of work, namely, the heat developed by compression, is entirely unavoidable. The whole of the mechanical energy which the compressor-piston spends upon the air is converted into heat. This heat is dissipated by conduction and radiation, and its mechanical equivalent is work lost. The compressed air, having again reached thermal equilibrium with the surrounding atmosphere, expands and does work in virtue of its *intrinsic energy*.

We proceed to the second loss, which is the work done in driving the compressor-piston against the increase of pressure due to the heat of compression. Since the temperature increases more rapidly than it ought, according to Boyle's law, the work necessary to compression is greater than if the temperature were to remain constant.

The theoretical efficiency of the compressing and working cylinders, as given further on by eq. (486), is:

$$E = \frac{\theta_0}{\tau_1},$$

where τ_1 is the absolute temperature of the air at its exit from the compressor, and θ_0 the absolute temperature at its entrance into the working cylinder, which in practice is that of the surrounding atmosphere. Hence we can increase the value of this fraction only by decreasing the denominator τ_1, that is the final heat of compression. This can only be done by abstracting the heat during compression, or by using very low pressures. But low pressures are excluded by other considerations. The weight of air, w, needed per second to perform a given amount of work would have to be considerably increased, and this would necessitate larger pipes, larger cylinders, and would result in a cumbrous and expensive arrangement.

The only remaining alternative, therefore, is to bring about in the compressor the cooling which the air now undergoes after having left it. Table VII shows respectively the portion of work lost when the air is not cooled in the compressor and that lost when it is completely cooled, and will make manifest the advantage there is in cooling. For a pressure of six atmospheres the work spent in isothermal compression to that spent in adiabatic compression is as 3 to 4; and this ratio decreases rapidly as the pressure increases.

II.

METHODS OF COOLING.

There are three methods in which cold water is applied to cool the air during its compression:

1. In case of the so-called hydraulic piston or plunger compressors, the air is over and in contact with a column of water which acts upon the air like an ordinary piston, its surface rising and falling with the backward and forward motion of the plunger. It is obvious that the cooling effect of this large mass of water is very small. There is nothing but surface contact, and water possesses in a slight degree only, the property of

conducting, through its mass, heat received on its surface. But we obtain all the advantages there are in having the air completely saturated with water-vapor during its compression, as well as all the disadvantages of having saturated compressed air to work with. What has been here said of hydraulic plunger-compressors, applies equally to hydraulic or ram compressors (first used by Sommeiller at Mt. Cenis, but now obsolete).

2. By flooding the external of the cylinder, and sometimes also the piston and piston-rod. This method of cooling presents neither the advantages nor disadvantages incident to direct intercontact between the air and water; it is that generally adopted in American practice, especially where it is necessary to expose the air-pipes to the out-door atmosphere of winter. The cooling which it effects is, however, only an approach to that which insures the highest efficiency.

3. By injecting into the compressor-cylinder a certain quantity of water in a state of the finest possible division, *i. e.*, in

the form of *spray*. This method of cooling was first applied by Prof. Collodon in the compressors used at the St. Gothard Tunnel. It is by far the most rational, complete and effective. In this fine state of division the water has many more points of contact with the air, which is both completely cooled and kept thoroughly saturated during compression. It is extremely important that the quantity of water injected into the compressor be a minimum, and hence the weight required for different tensions is given in a table further on.

III.

CONDITIONS MOST FAVORABLE TO ECONOMY IN THE USE OF COMPRESSED AIR.

By working air at full pressure we avoid the formation of ice in the pipes and exhaust ports, not so much because the air is less cooled (for the great fall of temperature produced by the sudden expansion at the instant of exhaust is almost equal to that produced by interior expansion), but because the air in exhausting acquires a high velocity, and this opposes the deposit of ice crystals by its purely mechanical effect, and by the heat developed by its friction.

But even at full pressure we cannot work with high tensions without serious drawbacks. In England, several trials were made at the Govan Iron Works and other places to use air under tensions of eight and nine atmospheres, but they were forced to return to low pressures, owing to the entire arrest of the machine from the formation of ice in the ports. Hence, not taking into account the fact that the useful effect decreases as the pressure increases, we conclude that it is not good practice, even at full pressure, to work with a tension much over four atmospheres, unless we employ special means to reheat the working air.

But while by working at full pressure with moderate tensions, we avoid the inconveniences of very low temperatures, the efficiency obtained is also very low. Notwithstanding this, even up to the present time air is almost exclusively worked at full pressure, especially in the United States. This is because the great cold produced by expansive working has made its adoption impossible. With a cut-off at $\frac{1}{2}$ stroke the temperature of the air falls 71° C, and at $\frac{1}{5}$ cut-off 140° C.

Now, to avoid these low temperatures, it is necessary either that the initial temperature of the compressed air be raised by heating it before its introduction into the working cylinder, or that the cylinder in which it expands be heated, or that the compressed air be supplied with heat directly during its expansion by means of the injection of hot water.

In 1860, M. Sommeiller, in order to utilize expansion, heated his working cylinders at Bardonnéche by means of a current of hot air circulating around the cylinders in small pipes. By this means he was enabled to cut off at $\frac{2}{5}$ stroke.

In 1863, M. Devillez recommended that the cylinder be placed in a tank through which hot water was to circulate. Other devices were to place the cylinder into a tank of water, into which from time to time fresh supplies of quicklime were to be thrown. Waste cotton, soaked in petroleum, was also used to heat the working cylinder.

Finally, in 1874, Mr. C. W. Siemens proposed the injection of hot water into the compressed air engine cylinder to keep the temperature of the expanding air from falling below the freezing point, just as we inject cold water into the compressor cylinder to prevent a great rise of temperature during compression. This is by far the most efficient mode of supplying heat to the expanding air. Expansion is made completely practicable, and hence the efficiency of the engine is greatly increased, as was shown by M. Cornet, who was the first to apply Mr. Siemens' plan and to prove conclusively its great practical utility.

The quantities of hot water to be injected into the cylinder should always be a minimum; they are given in a table further on.

IV.

EFFICIENCY ATTAINED IN PRACTICE.

It is desirable to know what efficiencies have been attained in practice—of compressors, of compressed-air engines, and of the two machines together as a system.

1. By efficiency of compressor is meant the ratio of the effective work spent upon the air in the compressor to that developed by the steam in the driving engine; or if you choose, the resistance divided by the power.

a. In compressors without piston or plunger, such as the hydraulic compressor of Sommeiller, the efficiency is always less than .50. These machines are interesting on account of their simplicity, but their useful effect is always very small.

b. In the so-called hydraulic piston or plunger compressor, an efficiency of .90 has been obtained when working at a low piston-speed to pressures of four and five atmospheres.

c. The compressors of Albert Schacht

at Saarbrücken, in which the cooling is wholly external, have shown an efficiency of .80 when compressing to a tension of 4 effective atmospheres.

d. Prof. Collodon's compressors, into which water is injected in the form of spray, and which were run at a piston-speed of 345 feet, and compressed the air to an absolute tension of 8 atmospheres, gave an efficiency which never descended below .80, while the temperature of the air never rose higher than from 12 to 15 degrees C.

2. The efficiency of compressed-air engines is the ratio of the work which they actually do, to that which is theoretically obtainable from the compressed air. The following are examples of its value as found by experiment:

At the Haigh Colliery, Eng., .70
"   " Ryhope "   "   .66

M. Ribourt has found for his locomotives .50 to .60.

In general it may be said that in the very best machines we can count upon from .70 to .75; while in the ordinary

ones, working against a variable resistance, this efficiency descends to .50 and .55.

3. The efficiency of the whole system together, that is, the ratio of the work measured on the crank-shaft of the compressed-air engine, to that done by the prime mover, is found to be about .20 to .25 for high pressures, and from .35 to .40 for low pressures.

Experiments made at Leeds show a net efficiency of .255 when working with 2.75 effective atmospheres, and .455 when with 1.33 atmospheres.

At the Blanzy mines, M. Graillot has found for a final efficiency, .22 to .32 of the effective work of the steam.

M. Ribourt, by experimenting on the new compressed-air locomotives built for the St. Gothard Tunnel, found that the ratio of the tractive effort developed to the original power, (in this case a head of water), was .23; that is, after passing the turbine, the compressor, the expansion regulator, and the cylinders of the locomotive, there remained .23 of the original power.

V.

THE EFFICIENCY OF FULL PRESSURE AND OF EXPANSION COMPARED.

Let W_1 be the work spent upon the air in the compressor;

W_2 the work which the compressed air is theoretically able to do; then its theoretical efficiency will be $\dfrac{W_2}{W_1}$.

If W = the actual work done by the prime mover, and

W' the actual work done by the air, then the real efficiency will be $\dfrac{W'}{W}$.

Now in the ordinary conditions of practice we know that W_1 is at best .70 W, and W' is only about .70 W_2; hence

$$E' = \text{real efficiency} = \frac{W'}{W} = \frac{.70 W_2}{\frac{W_1}{.70}} = .49 \frac{W_2}{W_1} = .49 E.$$

The value of $\dfrac{W_2}{W_1}$ ($= E =$ the theoretical efficiency) is .55 for full pressure and .75 for complete expansion. Hence, substituting these values of E above, we find for these two cases a final efficiency of .27 and .37.

VI.

LOSSES OF TRANSMISSION.

The losses due to transmission are calculated further on.

At the works for excavating the Mt. Cenis Tunnel, the supply of compressed air was conveyed in cast iron pipes $7\frac{5}{8}$ inches in diameter. The loss of pressure and leakage of air, from the supply pipes, in a length of *one mile and fifteen yards*, was only $3\frac{1}{2}\%$ of the head; the absolute initial pressure was 5.70 atmospheres, and it was reduced to 5.50 atmospheres whilst there was an expenditure at the rate of 64 cubic feet of compressed air per minute. In the middle of the tunnel, through a length of pipe of 3.8 miles, the absolute pressure fell only from six atmospheres to 5.7 atmospheres, or to .95 of the original pressure.

At the Hoosac Tunnel the air was carried through an 8-inch pipe from the compressors to the heading, a distance of

7,150 feet, operating six drills, with an average loss of *two* pounds pressure.

CHAPTER II.
The Physical Properties and Laws of Air.
I.
INTRODUCTORY.

A fluid is *a body incapable of resisting a change of shape.* Fluids are either liquids, vapors or gases. Water may be taken as the type of the first; steam is the type of all vapors, and air of all gases.

Gases are either *coercible* gases, *i. e.*, such as under ordinary circumstances may be condensed into liquids or even solids, as CO_2; or *permanent* gases, which retain their aëriform state under all ordinary circumstances of temperature and pressure. This distinction is convenient. Air has been condensed, but certainly not under *ordinary* circumstances.

Air then is a *permanent* gas, and may be considered a *perfect fluid;* that is,

1. It is incapable of experiencing a distorting or tangential stress, its molecules offering no resistance to relative displacement among themselves; hence no internal work of displacement need be considered.

2. It has the power of indefinite expansion so as to fill any vessel of whatever shape or size.

3. It exerts an equal pressure upon every point of the walls of the vessel enclosing it.

4. It is of the same density at every point of the space it occupies.

II.

BOYLE'S LAW.

This law states that *the temperature being constant, the volume of a gas varies inversely as the pressure;* formulated:

$$pv' = p_0 v_0 \qquad (1)$$

Where v_0 = the volume of a given weight of the gas at freezing temperature and a pressure p_0; and v' = the volume of the same weight of gas at the same temperature and at any pressure p.

Dry air, a mechanical mixture of oxygen and nitrogen, being a permament gas, obeys this law.

III.

THE LAW OF GAY-LUSSAC.

This second law of gases may be stated thus: *The volume of a gas under constant pressure expands when raised from the freezing to the boiling temperature, by the same fraction of itself, whatever be the nature of the gas; formulated:*

$$v = v^1 (1 + a_1 t) \qquad (2)$$

It has been found by the careful experiments of MM. Rudberg, Regnault and Prof. Balfour Stewart and others, that the volume of air at constant pressure expands from 1 to 1.3665 between 0° C. and 100° C. Hence for a variation in temperature of 1° C., the volume varies by .003665 or $\frac{1}{273}$ of the volume which the air occupied at 0° C. and under the assumed constant pressure. In equation (2) the coefficient a_1 is therefore equal to $\frac{1}{273}$.

IV.

BOYLE'S AND GAY-LUSSAC'S LAW.

Combining the equation formulating Boyle's law with that formulating Gay-Lussac's, we obtain,

$$pv = p_0 v_0 (1 + a_1 t) = p_0 v_0 a_1 \left(\frac{1}{a_1} + t\right);$$

or letting $a = \dfrac{1}{a_1} = 273$, we have

$$pv = \frac{p_0 v_0}{a}(a+t) = R(a+t) \qquad (3)$$

This last equation is a general expression for both Boyle's and Gay-Lussac's law, and completely expresses the relation between temperature, volume, and pressure.

R is a constant and depends upon the density of the gas. Its value for atmospheric air is determined as follows:

The weight of the standard unit of volume of a substance in any condition is the *specific weight* of that substance in that condition.

The *specific weight* of air, that is t say, the weight of a cubic foot of air at

0° C. and under a pressure of 29.92 inches of mercury, is, according to M. Regnault, .080728 lbs. avoirdupois.

The *specific volume* of a gas is the volume of unit of weight; it is the reciprocal of the specific weight.

The *specific volume* of air, *i.e.*, the volume in cubic feet of one pound avoirdupois at 0° C. and under the pressure of 29.92 inches mercury is:

$$v_0 = \frac{1}{.080728} = 12.387 \text{ cubic feet.}$$

Let $p_0 = 2116.4$, the mean atmospheric pressure in lbs. per square foot. Then

$$R = \frac{p_0 v_0}{a} = \frac{2116.4 \times 12.387}{273} = 96.0376.$$

V.

ABSOLUTE TEMPERATURE.

Making $t = -273$ in the equation
$$pv = R(a + t),$$
the second member reduces to zero, and hence,
$$pv = o.$$

The distance of the freezing point from the bottom of the tube of an air thermometer is to the distance of the boiling point from the bottom as $1:1.3665$. Hence, in the centigrade scale, where the freezing point is marked $0°$ and the boiling point $100°$, the bottom of the tube will be marked—$272°.85$. The lowest reading of the scale is, therefore, $-273°$. If this reading could be observed it would imply that the volume of the air had been reduced to nothing. This is evidently a purely theoretical conception; but in dealing with questions relating to gases it is exceedingly convenient to reckon temperatures, not from the freezing point, but from the bottom of the tube of an air thermometer. *Absolute zero*, therefore, is marked $-273°$ on the Centigrade scale (corresponding to $-459.°4$ on Fahranheit's scale), and is the temperature at which all molecular motions cease, and the mechanical effect, which we call pressure, and which is due to these motions, becomes zero.

VI.

LAW OF THE PRESSURE, DENSITY AND TEMPERATURE.

Let $D_0 =$ the density of a weight w of air at the temperature $0°$ C. and under the pressure p_0, v_0 being the corresponding volume;

$D =$ its density at pressure p, temperature t, v being its corresponding volume;

$D' =$ its density at temperature $0°$ C. pressure p and volume v'.

We shall have

$$D = \frac{w}{v},$$

or by taking $w =$ unity,

$$D = \frac{1}{v'}, \text{ and } v = \frac{1}{D}.$$

Placing these values of v' and v_0 in equation (1), we get

$$\frac{p}{p_0} = \frac{D'}{D_0}; \qquad (4)$$

that is, *the pressure of a gas is proportional to its density.*

From (2) we have,

$$\frac{D}{D'} = \frac{1}{1 + a't} = \frac{a}{a + t}; \qquad (5)$$

That is, the density of a gas is inversely as its temperature, the latter being reconed from absolute zero.

Combining equations (4) and (5),

$$\frac{D}{D_0} = \frac{p}{p_0} \times \frac{a}{a+t}, \text{ or}$$

$$p = \frac{p_0}{D_0} \times \frac{a+t}{a} D. \qquad (6)$$

But $D = \frac{w}{v}$, and hence

$$pv = \frac{p_0}{D_0} \times \frac{(a+t)}{a} w. \qquad (6a)$$

(6) shows that the density of a gas is:

At constant temperature, directly as the pressure;

At constant pressure, inversely as the absolute temperature.

$\frac{p_0}{D_0}$ = constant for any given gas. For air $\frac{p_0}{D_0} = \frac{2116.4}{.080728} = 26216.43$ (according to Rankine, 26214); this is the height in feet of a column of fluid of density D_0, which produces a pressure p_0 pounds per square foot of surface; letting H be this height, the weight of the column having one square foot for its surface will be $D_0 H$, or

$$D_0 H = p_0.$$

If in (6a) we make $v = 1$, we get

$$w = \frac{p}{a+t} \times \frac{D a}{p_0} = \frac{p}{a \times t} \times \frac{1}{R} \quad (7)$$

which is the weight of unit of volume, or the *specific weight* of air.

Making $w = 1$ in same equation, we have for the volume of unit of weight,

$$v = \frac{p_0}{D_0 a} \times \frac{a+t}{p} = R \frac{a+t}{p} \quad (8)$$

called the *specific volume*. (7) and (8) are reciprocals of each other.

VII.

THE MEASUREMENT OF HEAT.

Any effect of heat may be used as a means of measuring it, and the quantity of heat required to produce a particular effect is called a thermal unit. It has been found best to take a thermal unit to be the quantity of heat which corresponds to some definite interval of temperature in a definite weight of a particular substance.

Def. A *British Thermal Unit* is the quantity of heat which corresponds to an interval of one degree of Fahrenheit's scale, in the temperature of one pound of pure liquid water at its temperature of greatest density (39°.1 Fahr).

Def. A *Calorie*, or French Thermal Unit, is the quantity of heat which corresponds to the *Centigrade* degree in the temperature of one *kilogram* of pure liquid water, at its temperature of greatest density, (3°.94 C.).

Def. The *Specific Heat* of a body, is the ratio of the quantity of heat required to raise that body one degree, to the quantity required to raise an equal weight of water one degree.

It has been proven for permanent gases, that,

1. The specific heat is constant for any given gas, and is independent of the temperature and pressure;

2. The thermal capacity per unit of

volume, is the same for all simple gases when at the same pressure and temperature;

3. The specific heat increases with the temperature, and probably with the pressure, when the gas is brought near the point of liquifaction, and no longer obeys Boyle's law.

The above three conclusions are true of specific heat at *constant volume*, as well as of specific heat at *constant pressure*, as far as regards simple gases and air, (which, being a mechanical mixture, obeys the same laws as a simple gas).

It was shown by Laplace, that the specific heat of a gas is different, according as it is maintained at a constant *volume*, or at a constant *pressure*, during the operation of changing its temperature.

The specific heat of gases was independently determined by M. Regnault and Prof. Rankine; experimentally by the former, and theoretically by the latter. Their results agreed exactly, and are those now generally accepted.

As given in Watt's Dictionary of Chemistry,

The specific heat at constant pressure is .238.

As we shall find farther on, the specific heat at constant volume is .169.

$$\therefore \frac{c}{c^1} = \frac{.238}{.169} = 1.40 = \gamma.$$

CHAPTER III.

THERMODYNAMIC PRINCIPLES AND FORMULAS.

I

INTRODUCTORY.

It is well known that the cylinder of an air compressor becomes very hot even at a low piston-speed. This fact brings us face to face with the doctrine of the conversion of energy; for it is the conversion of the visible, mechanical energy of the piston into that other invisible form of energy called heat. Thus we see we are at the very outset confronted with a thermal phenomenon, whose consideration involves the science called thermodynamics. To begin with we had no other but the visible mechanical energy of a moving piston; but very soon sensible heat manifests itself, and this heat can be developed only at the expense of part at least, of the energy of the moving piston.

These phenomena are referable to the two general principles which form the basis of the science of thermodynamics, viz :

1. All forms of energy are convertible.

2. The total energy of a substance or system cannot be altered by the mutual actions of its parts.

"The conversion of one form of energy into another takes place with as great certainty and absence of waste, and with the same integrity of the elementary magnitude, as the more formal conversion of foot-pounds in kilogrammeters." "In the development of the axioms that nothing is by natural means creatable from nothing, and that things are equal to the same thing only which are equal to each other, and in the application to them of empirical laws with reference to the behavior of bodies under the action of heat and mechanical effect"* consists chiefly the science of thermodynamics.

The general equation of thermodymamics which expresses the relation between heat and mechanical energy under all circumstances, was arrived at independently, in 1849, by Professors Clausius and Rankine. The consequences of that equation have since been developed and applied by many distinguished writers.

Of course we shall here confine ourselves to so much only of the Mechanical Theory of Heat as is necessary to an intelligent comprehension of our subject; and, in doing so, shall follow in outline the treatment given by M. Pochet, in his admirable "*Nouvelle Méchanique Industrielle*," making free use, at the same time, of the works of Zeuner, Rankine and Clausius.

---

* "History of Dymamical Theory of Heat," by the late Porter Poinier, M.E., in *Popular Science Monthly* for January, 1878.

II.

HEAT AND TEMPERATURE.

Heat denotes a motion of particles on a small scale just as the rushing together of a stone and the earth denotes a motion on a large scale, a mass motion. It is due to a vibratory motion impressed upon the molecules of a body. The more rapid the vibrations the more intense the heat. The quantity of heat in a substance could be measured by multiplying the kinetic energy of agitation of a single molecule by the number of molecules in unity of weight, supposing the substance to be homogeneous and the heat uniformly distributed. Thus the thermometer and dynamometer reveal to us phenomena which are in reality identical, and we can establish a measuring unit to which both effects can be referred.

Temperature is the property of a body considered with reference to its power of heating other bodies. It is a function of the variables, volume and pressure, or,

$$t = \Phi(v, p);$$

that is, *all bodies having the same pressure and volume have the same temperature.* This is expressed by the differential equation:

$$dt = \left(\frac{dt}{dp}\right)dp + \left(\frac{dt}{dv}\right)dv, \qquad (9)$$

where $\left(\frac{dt}{dp}\right)$ and $\left(\frac{dt}{dv}\right)$ are the partial differential co-efficients; dt in the former denoting the increment of t when, v remaining constant, p alone is increased by dp; and in the latter, the increment received by t when p remaining constant, v is increased by dv; whilst in the first member of the equation, dt represents the *total* increment of t due to the simultaneous reception by p and v of the increments dp and dv, respectively.

III.

THE TWO LAWS OF THERMODYNAMICS.

The whole mechanical theory of heat rests on two fundamental theorems: *

1. *That of the equivalence of heat and*

---

* See Clausius on Heat, Memoir.

work; whensoever a body changes its state in producing exterior work, (positive or negative), there is an absorption or disengagement of heat in the proportion of one British thermal unit for every 772 foot pounds of work, (or of one French thermal unit for every 423.55 kilogrammeters of work).

This mechanical equivalent of heat was first *exactly* determined by Mr. Joule, in honor of whom it is called Joule's equivalent, and is denoted by the symbol J.

2. *The theorem of the equivalence of transformations;* when a body is successively put in communication with two sources of heat, one at a higher temperature t, the other at a lower temperature t_0, its temperature remaining constant and equal to that of each source during the whole time of contact, and the body neither receiving nor losing heat except by reason of its contact with the two sources, the ratio of the quantity of heat Q given out by the higher source to the quantity Q' transferred to the lower

source, is independent of the nature of the bodies; it depends only on the temperatures, t and t_0, of the two sources.

Clausius states this as follows: In all cases where a quantity of heat is converted into work, and where the body effecting this transformation ultimately returns to its original condition, another quantity of heat must necessarily be transferred from a warmer to a colder body; and the magnitude of the last quantity of heat, in relation to the first, depends only on the temperature of the bodies between which heat passes, and not upon the nature of the body effecting this transformation; or, more briefly, heat cannot *of itself* pass from a colder to a warmer body.

IV.

HEAT AND MECHANICAL ENERGY.

The quantity of heat which must be imparted to a body during its passage, in a given manner, from one condition to another, (any heat withdrawn from the body being counted an important negative quantity) may be divided into three parts, viz:

1. That employed in increasing the heat actually existing in the body;
2. That employed in producing interior work.
3. That employed in producing exterior work.

The first and second parts, called respectively the *thermal* and *ergonal* content* of the body, are independent of the path pursued in the passage of the body from one state to another; hence both parts may be represented by one function, which we know to be completely determined by the initial and final states of the body. The third part, the equivalent of exterior work, can only be determined when the precise manner in which the changes of condition took place is known.

Let $dQ=$ the element of heat absorbed during an infinitesimal change of condition;

---

* Clausius on Heat. Memoir.

U_0 = the free heat present in the body at the beginning, *i.e.*, the body's intrinsic energy;

U = the free heat present in the body at the end of the change, plus the heat consumed by internal work during the change of state;

pdv will be the work accompanying the passage of the body from a state (p, v) to a state $(p+dp, v+dv)$;

Then the heat spent while the body passes from one temperature t to another $t+dt$, and from one state (p, v) to another $(p+dp, v+dv)$ will be:

$$dQ = (U - U_0) + \frac{1}{J}.pdv,$$

$$= dU + \frac{1}{J}.pdv; \qquad (10)$$

where dU depends upon the *initial* and *final* circumstances, while $\frac{1}{J}.pdv$ depends on the intermediate circumstances of the change of state.

We can write $dU = o$ and entirely exclude interior work and heat by confining ourselves to *cyclical processes*, that is to say, to operations in which the modifications which the body undergoes are so arranged that the body finally returns exactly to its original condition, the interior work, positive and negative, exactly neutralizing each other.

$$U = f(p, v),$$

that is, the internal heat of a body depends only upon the volume of the body, and the pressure to which it is subjected. Hence the increase of internal heat when the body passes from a state (p, v) to a state $(p+dp, v+dv)$ will be:

$$dU = \left(\frac{dU}{dp}\right)dp + \left(\frac{dU}{dv}\right)dv. \quad (11)$$

Substituting in equation (10) the value of dU as given by equation (11), we have

$$dQ = \left(\frac{dU}{dp}\right)dp + \left\{\left(\frac{dU}{dv}\right) + \frac{p}{J}\right\}dv \quad (12)$$

an equation which is not integrable; since this would require that the second derivatives of the co-efficients of dp and dv (which are, respectively, $\frac{d^2U}{dp.dv}$ and

$\frac{d^2U}{dv.dp} + J$) should be equal to each other*; this would imply the impossible condition $J = o$. That is, mechanically speaking, the quantity of heat passing cannot be expressed as a function of the initial values of p and v. The equation can only be integrated when we have a relation given, by means of which t may be expressed as a function of v, and therefore p as a function of v alone. It is this relation which defines the manner in which the changes of condition take place; the quantity of heat passing depends upon the *intermediate* circumstances of change of state, circumstances which may be anything.

When a body is heated from a temperature t to another $t + dt$, preserving the *same volume*, no external work will be done and $dv = o$. Hence eq. (12) will become:

$$dQ = \left(\frac{dU}{dp}\right) dp$$
$$= c_1\, dt \qquad (13)$$

---

* See Ray's Infinitesimal Calculus, p. 366; also McCullough on Heat, arts. 61 and 62.

which, by definition, is the *specific heat at constant volume*.

The above equation gives:

$$\frac{dU}{dp} = c_1\left(\frac{dt}{dp}\right) \qquad (13a)$$

the partial differential co-efficient of t with respect to p.

If the body passes from t to $t+dt$ under *constant pressure*, $dp=0$, and hence (12) becomes:

$$dQ = \left\{\left(\frac{dU}{dv}\right) + \frac{p}{J}\right\} dv = c\, dt \qquad (14)$$

which, by definition, is the *specific heat at constant pressure*.

From (14) we have:

$$\left(\frac{dU}{dv}\right) + \frac{p}{J} = c\left(\frac{dt}{dv}\right). \qquad (14a)$$

Substituting these values of the partial derivatives in eq. (12), we obtain a second expression for dQ, viz.:

$$dQ = c_1\left(\frac{dt}{dp}\right)dp + c\left(\frac{dt}{dv}\right)dv \qquad (15)$$

It is convenient to have this equation in a form involving only the temperature and specific heats, and not the quantity Q. We obtain such a form by differentiating (13a) with respect to v, and (14a) with respect to p and subtracting the first result from the second. The form obtained is:

$$\frac{1}{J} = (c-c_1)\frac{d^2t}{dv.dp} + \left(\frac{dc}{dp}\right)\left(\frac{dt}{dv}\right) - \left(\frac{dc_1}{dv}\right)\left(\frac{dt}{dp}\right) \qquad (16)$$

V.

THE DIFFERENTIAL EQUATION OF THE SECOND PRINCIPLE.

In the figure,

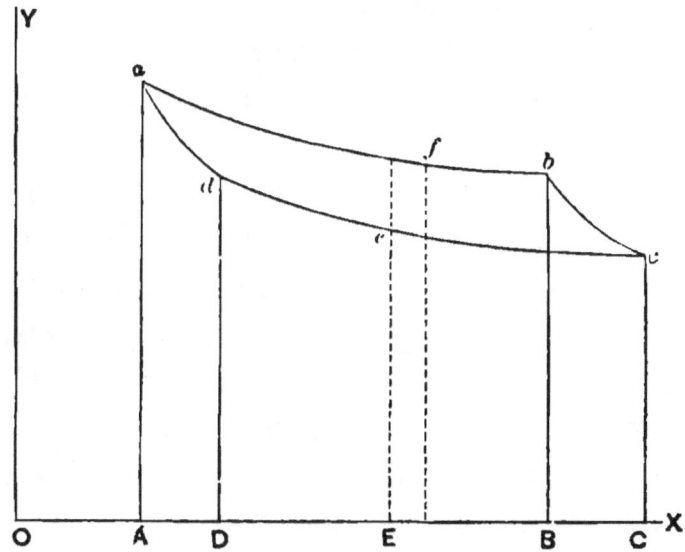

1. Let OA = the initial volume of a body whose temperature is t; it expands in contact with a source of heat, (isothermally), from volume OA to volume OB, when its temperature is then still t.

Q = the quantity of heat supplied by the source.

2. It is now left to expand adiabatically, *i.e.*, without the addition or subtraction of heat, from volume OB to volume OC, when its temperature will have fallen to t_0.

3. Now place it in contact with a source of heat of the same temperature t_0, and compress it from OC to OD, when its temperature is still t_0.

Q' = the quantity of heat that has passed into the source.

4. Compress it adiabatically from volume OD to volume OA, when its temperature will again be t; the body has now undergone a *complete cycle*, during which it has evidently done work represented the area *abcd*; hence,

Q − Q' = heat disappeared, and from the first law of thermodynamics,

$$Q - Q_1 = \frac{1}{J} \times abcd = \frac{1}{J} \times A. \qquad (17)$$

Now the second law of thermodynamics states that Q and Q', (the heat received and the heat given out), are independent of the nature of the bodies, and dependent only upon the temperature.

Suppose that the difference of temperature of the two sources of heat is infinitely small, t and $t+dt$. Also consider t and v as the independent variables determining the state of the body, $p = f(v, t)$.

A, in the above equation, is the integral between v_0 and v of the elementary areas, such as ef. Now if Ee=p, Ef is what p will become when the volume remains constant, and the temperature takes an increment dt; fe therefore measures the differential increment

$$\left(\frac{dp}{dt}\right)dt,$$

where $\frac{dp}{dt}$ = the partial derivative of p with respect to t.

Hence, $Q - Q' = A \cdot \dfrac{1}{J} = \dfrac{1}{J} \int_{v_0}^{v} \left(\dfrac{dp}{dt}\right) dt\, dv$

$\qquad\qquad = \dfrac{1}{J} dt \int_{v_0}^{v} \left(\dfrac{dp}{dt}\right) dv,$

taking the independent variable dt out of the integration symbol.

Q is the heat supplied to keep at t the temperature of the body expanding from v_0 to v; and, therefore,

$Q = \Phi(t, v_0, v$; the nature of the bodies);

also,
$$Q' = F''(t, v_0, v) = F(t),$$
the variables v_0, v being implicitly contained in F.

Since $Q = Q'$ when t becomes $t + dt$ we have,
$$Q = F(t + dt) = F(t) + F'(t)\, dt$$
and
$$\dfrac{Q}{Q_1} = 1 + \dfrac{F'(t)}{F(t)} dt.$$

According to the second principle, $\dfrac{Q}{Q'}$ is independent of the nature of the bodies; hence,

$$Q - Q' = Q' \dfrac{F'(t)}{F(t)} dt = \dfrac{1}{J} dt \int_{v_0}^{v} \left(\dfrac{dp}{dt}\right) dv$$

and

$$Q' = \frac{1}{J} \frac{F(t)}{F'(t)} \int_{v_0}^{v} \left(\frac{dp}{dt}\right) dv.$$

Now, suppose $v - v_0$ becomes indefinitely small and equal to dv; Q' will become dQ, Q being the heat necessary to keep at t the temperature of a body whose volume increases by dv; hence the differential equation of the first order,

$$dQ = \frac{1}{J} \Phi(t) \frac{dp}{dt} dv \qquad (18)$$

is the differential equation of the second principle.*

Calculation of the function $\Phi(t)$. It may have several forms. Making $dt = o$ in eq. (9), we get,

$$dp = -\frac{\left(\dfrac{dt}{dv}\right)}{\left(\dfrac{dt}{dp}\right)} dv;$$

Placing this value of dp in eq. (15),

$$dQ = (c - c_1) \left(\frac{dt}{dv}\right) dv.$$

---

* See Zeuner, "Théorie Méchanique de la Chaleur," troisième section, iii.
Also, Clausius on Heat, first Memoir.

Moreover in (9) $\left(\dfrac{dp}{dt}\right)$ represents the partial derivative of p with relation to t when v is constant; making $dv = o$,

$$\left(\dfrac{dp}{dt}\right) = \dfrac{1}{\left(\dfrac{dt}{dp}\right)};$$

Hence eq. (18) may be written,

$$dQ = \dfrac{1}{J}\Phi(t)\dfrac{dv}{\left(\dfrac{dt}{dp}\right)}.$$

Equating this with the value of dQ above, we have,

$$\dfrac{1}{J}\Phi(t) = (c-c_1)\left(\dfrac{dt}{dp}\right)\left(\dfrac{dt}{dv}\right), \quad (19)$$

from which $\Phi(t)$ may be calculated.

Again, if we take Eq. (16) and suppose it applied to bodies whose specific heats c and c_1 are independent, the first of the pressure and the second of the volume, as is the case in permanent gases, these conditions give $\left(\dfrac{dc}{dp}\right)$ and $\left(\dfrac{dc_1}{dv}\right)$ equal to zero, and the equation becomes,

$$(c-c_1)\left(\frac{d^2t}{dp\,dv}\right)=\frac{1}{\mathrm{J}}. \qquad (20)$$

Dividing eq. (19) by this we get,

$$\Phi(t)=\frac{\left(\dfrac{dt}{dp}\right)\left(\dfrac{dt}{dv}\right)}{\dfrac{d^2t}{dp\,dv}} \qquad (21)$$

giving $\Phi(t)$ as a function of $t\ [=f(p,v)]$ and of its partial derivatives.

CHAPTER IV.

THE THERMODYNAMIC EQUATIONS APPLIED TO PERMANENT GASES.

I.

DETERMINATION OF THE SPECIFIC HEAT AT CONSTANT VOLUME.

Forming, from eq. (3), the partial differentials:

$$\left(\frac{dt}{dp}\right)=\frac{v}{\mathrm{R}},\ \left(\frac{dt}{dv}\right)=\frac{p}{\mathrm{R}},\ \frac{d^2t}{dp.dv}=\frac{1}{\mathrm{R}},$$

and substituting in eqs. (20) and (21), we have:

$$(c-c')=\frac{1}{\mathrm{J}}\mathrm{R}, \qquad (22)$$

and $\Phi(t) = \dfrac{pv}{R} = (a+t).$ (23)

(22) gives, $c' = c - \dfrac{1}{J}R = .238 - \dfrac{96.0376}{1389.6}$
$= .169$

which is the specific heat at constant volume for atmospheric air.

II.

INTERNAL HEAT.

Placing eqs. (12) and (15) equal to each other and substituting the value of c from (22), we have:

$$\left(\frac{dU}{dp}\right)dp + \left(\frac{dU}{dv}\right)dv = c'(vdp + pdv)\frac{1}{R} = dU$$

according to eq. (11).

Integrating, and substituting for R its value $\dfrac{pv}{\tau}$ we have,

$$U = c'\tau - U_0$$
or $\qquad U - U_0 = c'\tau \qquad$ (24)

which shows that the internal heat for every degree of temperature is increased by a quantity c' (.169), and the increase of the internal heat of a gas passing from 0°C. to t°C. is always the same, whatever variations its pressure may have undergone in this passage, the volume having been kept constant.

III.

QUANTITY OF HEAT SUPPLIED.

The partial differentials formed from eq. (3) placed in (15) gives:

$$dQ = \frac{c'vdp + cpdv}{R} \qquad (25)$$

which is integrable only when we have a given relation between p and v.

1. At constant volume; make $dv = o$, v being constant. Then

$$Q = \int_{p_0}^{p} \frac{c'vdp}{R} = \frac{c'v(p-p_0)}{R} = c'(\tau - \tau_0) \qquad (25a)$$

which defines the specific heat at constant volume.

2. At constant pressure; here $dp = o$, and eq. (25) gives:

$$Q = \frac{cp(v-v_0)}{R} = c(\tau - \tau_0) \qquad (25b)$$

IV.

EXPANSION AT A CONSTANT TEMPERATURE.

To find the work done by a gas expanding *isothermally*, (that is, the absolute temperature is maintained at a constant value), we must satisfy Boyle's law and write:

$$pv = p_0 v_0 = \text{constant};$$

hence $pdv + vdp = 0;$ or, $vdp = -pdv.$

Substituting this in (25),

$$dQ = \frac{(c-c')pdv}{R} = \frac{1}{J}pdv;$$

Introducing p from eq. (3),

$$dQ = \frac{1}{J}R(a+t)\frac{dv}{v},$$

and,

$$Q = \frac{1}{J}R(a+t)\int_{v_0}^{v} \frac{dv}{v} = \frac{1}{J}R(a+t)\log.\frac{v}{v_0}.$$

$$= \frac{1}{J}p_0 v_0 \log.\frac{v}{v_0} \qquad (26.)$$

Let W = the work done; then

$$W = p_0 v_0 \log.^* \frac{v\mathrm{I}}{v_0}. \qquad (26a.)$$

the ordinary form for permanent gases.

V.

EXPANSION IN A PERFECTLY NON-CONDUCTING CYLINDER.

If a gas expand *adiabatically*, (*i.e.*, without any passage of heat either into the gas from without or out of the gas into other substances), $dQ = o$ in eq. (25), and we have,

$$c'vdp + cpdv = o.$$

Writing for $\frac{c}{c'}$ its value γ, and integrating, we have

$$\int_{p_0}^{p} \frac{dp}{p} + \gamma \int_{v_0}^{v} \frac{dv}{v} = \log.\frac{p}{p_0} + \gamma \log.\frac{v}{v_0} = o$$

$$= \log.\frac{p}{p_0} + \log.\frac{v^\gamma}{v_0^\gamma} = o$$

or, $\log.\frac{p}{p_0} = \log.\frac{v}{v_0} \times (-\gamma) = \log.\frac{v_0^\gamma}{v^\gamma}$

---

* The logarithms, it is seen, are taken in the Naperian system.

hence, $pv^\gamma = p_0 v_0^\gamma =$ constant; (27)

an equation which expresses the variation of pressure as a function of volume when the expansion or compression is *adiabatic*.

The external work performed during a finite expansion is denoted by

$$W = \int_{v_0}^{v} p\, dv = \int_{v_0}^{v} p_0 v_0^\gamma \frac{dv}{v^\gamma} =$$

$$p_0 v_0^\gamma \int_{v_0}^{v} v^{-\gamma}\, dv \quad (27a)$$

$$= \frac{p_0 v_0^\gamma}{\gamma-1}\left(\frac{1}{v_0^{\gamma-1}} - \frac{1}{v^{\gamma-1}}\right) = \frac{p_0 v_0}{\gamma-1}$$

$$\left\{ 1 - \left(\frac{v_0}{v}\right)^{\gamma-1} \right\} \quad (28)$$

Since no heat is received from without, the thermal equivalent of the work must be estimated as *internal* heat. If, now, τ_0 and τ are the initial and final absolute temperatures, the decrease in internal heat will be

$$c'(\tau_0 - \tau).$$

Hence we must have,

$$c'(\tau_0 - \tau) = \frac{1}{J} \frac{p_0 v_0}{\gamma - 1} \left\{ 1 - \left(\frac{v_0}{v}\right)^{\gamma-1} \right\} \quad (29)$$

Eq. (27) gives $\frac{pv^\gamma}{p_0 v_0^\gamma} = 1$; multiplying both members by $\frac{v_0^{\gamma-1}}{v^{\gamma-1}}$ we have,

$$\frac{pv}{p_0 v_0} = \left(\frac{v}{v_0}\right)^{\gamma-1} = \frac{a+t}{a+t_0} = \frac{\tau}{\tau_0}; \quad (30)$$

also,

$$\frac{v_0^\gamma}{v^\gamma} = \frac{p}{p_0}, \text{ and } \frac{v_0}{v} = \left(\frac{p}{p_0}\right)^{\frac{1}{\gamma}}.$$

hence,

$$\left(\frac{v_0}{v}\right)^{\gamma-1} = \left(\frac{p}{p_0}\right)^{\frac{\gamma-1}{\gamma}} = \frac{a+t}{a+t_0} = \frac{\tau}{\tau_0} \quad (31)$$

Substituting in (28) the values of $p_0 v_0$ from (3) and $\left(\frac{v_0}{v}\right)^{\gamma-1}$ from (31), we obtain:

$$W = \frac{R(a+t_0)}{\gamma - 1} \left\{ 1 - \left(\frac{p}{p_0}\right)^{\frac{\gamma-1}{\gamma}} \right\} \quad (32)$$

a form often used . $\frac{\gamma-1}{\gamma} = .2908.$

VI.

VARIATIONS IN THE TEMPERATURE OF A GAS DURING EXPANSION OR COMPRESSION IN A PERFECTLY NON-CONDUCTING CYLINDER.

Placing the second members of $p_0 v_0 = R(a+t_0)$, $\gamma = \dfrac{c}{c'}$, and $J = \dfrac{c-c'}{R}$ in eq. (29) we get:

$$t_0 - t = \tau \left\{ 1 - \left(\frac{v_0}{v}\right)^{\gamma-1} \right\}, \qquad (33)$$

which is thus interpreted:

The decrease in temperature (during an expansion from v_0 to v) *is proportional to the initial absolute temperature.*

The already established relation,

$$\frac{\tau}{\tau_0} = \left(\frac{v_0}{v}\right)^{\gamma-1}$$

expresses the final temperature as a function of the volumes; and if we know the initial and final pressures, the final temperature is expressed as a function of these pressures as follows:

$$\frac{a+t}{a+t_0} = \frac{\tau}{\tau_0} = \left\{ \frac{p}{p_0} \right\}^{\frac{\gamma-1}{\gamma}}$$

CHAPTER V.

THERMODYNAMIC LAWS APPLIED TO THE ACTION OF COMPRESSED AIR.*

I

FUNDAMENTAL FORMULAS.

The four equations formulating the law for the expansion and compression of dry air, are, as we have established them,

$$\frac{pv}{a+t} = Rw = \frac{pv}{\tau} = J(c-c')w \qquad (34a)$$

$$\frac{p}{p_0} = \left\{\frac{v_0}{v}\right\}^{\gamma} = \left\{\frac{v_0}{v}\right\}^{\frac{c}{c'}} \qquad (34\,b)$$

$$\frac{\tau}{\tau_0} = \left\{\frac{v_0}{v}\right\}^{\gamma-1} = \left\{\frac{v_0}{v}\right\}^{\frac{c}{c'}-1} \qquad (34c)$$

$$\frac{\tau}{\tau_0} = \left\{\frac{p}{p_0}\right\}^{\frac{\gamma-1}{\gamma}} = \left\{\frac{p}{p_0}\right\}^{\frac{\frac{c}{c'}-1}{\frac{c}{c'}}} \qquad (34d)$$

---

* The subject of this chapter is very ably treated by M. Mallard, in the "Bulletin de la Société de l' industrie minerale," tome xii, page 615, to whom the writer is greatly indebted.

These expressions sum up the relations existing between the *pressure, volume* and *absolute temperature* of a weight of air w compressed or expanded in a perfectly non-conducting cylinder.

p_0, τ_0, and v_0 have reference to the *initial* state of the weight of air considered, p, τ and v corresponding to the final state.

The following table is that of MM. Mallard and Pernolet. It gives for convenient values of $\dfrac{p}{p_0}$ the corresponding values of $\dfrac{\tau}{\tau_0}$, &c. The tabular differences facilitate interpolation.

(*See Table on pages* 72 *and* 73.)

II.

WORK SPENT IN COMPRESSING AIR.

The compressing-cylinder being supposed perfectly non-conducting as to heat, our machine may be called a "Reversible Engine;" for by reversing the process of compression under exact-

.ly the same conditions, we get back the exact amount of work spent in the compression.

The net work necessary to compress a weight of air w, taken from a reservoir (as the atmosphere) in which the pressure p_0 is kept constant, and to force it into another reservoir in which the pressure is constantly p_1, is made up of the following parts:—

1. The work of compression:
2. Diminished by the work due to the pressure p_0 of the first reservoir (the atmosphere); this work is $p_0 v_0$, v_0 being the volume of weight w under pressure p_0 and at the temperature t_0:
3. Increased by the work necessary to force the compressed air into the receiving reservoir; this is given by the expression $p_1 v_1$, v_1 being the volume of a weight of air w at the pressure p_1 and temperature t_1.

As no heat passes between the air and external bodies, the thermal equivalent of the work, according to the mechanical theory of heat, is the difference between

Table I.

$\dfrac{p}{p_0}$	$\dfrac{\tau}{\tau_0}$		$\dfrac{\tau_0}{\tau}$		$1-\dfrac{\tau_0}{\tau}$
	Numbers.	Differences.	Numbers.	Differences.	Numbers.
1.2	1.0543	481	.9485	415	.0515
1.4	1.1024	436	.9070	344	.0930
1.6	1.1416	439	.8762	293	.1274
1.8	1.1859	367	.8433	254	.1567
2	1.2226	343	.8179	223	.1821
2.2	1.2569	321	.7956	198	.2044
2.4	1.2890	303	.7758	178	.2242
2.6	1.3193	187	.7580	161	.2420
2.8	1.3480	272	.7419	147	.2581
3	1.3752	260	.7272	124	.2728
3.2	1.4012	248	.7138	125	.2862
3.4	1.4260	238	.7013	116	2987
3.6	1.4498	230	.6897	107	.3103
3.8	1.4728	220	.6790	100	.8210
4	1.4948	213	.6690	94	.3310
4.2	1.5161	206	.6596	89	.3404
4.4	1.5367	200	.6507	81	.3493
4.6	1.5567	193	.6424	79	.3576
4.8	1.5760	188	.6345	75	.3655
5	1.5948	865	.6270	322	.3730
6	1.6813	769	.5948	260	4052
7	1.7582	694	.5684	217	.4512
8	1.8276	636	.5471	183	.4529
9	1.8712	588	.5288	159	.4712
10	1.9500	544	.5128	141	.4871
11	2.0044	512	.4988	124	.5012
12	2.0556	484	.4864	111	.5136
13	2.1040	457	.4753	101	.5247
14	2.1497	434	.4652	92	.5348
15	2.1931		.4560		.5440

Table I.—*Continued.*

$\dfrac{v_0}{v}$		$\dfrac{v}{v_0}$		$\dfrac{t_a}{t}$	
Numbers.	Differences.	Numbers.	Differences.	Numbers.	Differences.
1.1382	1317	.8786	911	.793	78
1.2699	1262	.7875	712	.695	53
1.3961	1218	.7163	575	.642	39
1.5179	1179	.6588	475	.603	32
1.6358	1145	.6113	400	.571	25
1.7503	1116	.5713	342	.546	22
1.8619	1088	.5371	297	.524	19
1.9707	1065	.5074	260	.505	17
2.0772	1043	.4814	230	.488	15
2.1815	1023	.4584	205	.473	13
2.2838	1005	.4379	185	.460	12
2.3843	587	.4194	167	.448	10
2.4830	972	.4027	151	.438	10
2.5802	957	.3876	139	.428	9
2.6759	943	.3737	127	.419	9
2.7702	930	.3610	117	.410	8
2.8632	118	.3493	111	.402	7
2.9550	906	.3384	101	.395	7
3.0456	896	.3283	93	.388	6
3.1352	4333	.3190	388	.382	27
3.5685	4129	.2802	290	.355	19
3.9814	3858	.2512	227	.334	19
4.3772	3817	.2285	184	.317	14
4.7589	3697	.2101	151	.303	12
5.1286	3583	.1950	126	.291	10
5.4869	3484	.1824	111	.281	9
5.8353	3430	.1713	95	.272	9
6.1783	334	.1618	83	.263	7
6.5123	3273	.1535	73	.256	6
6.8396		.1462		.250	

the quantity of internal heat possessed by the air at its entrance into the cylinder, and that possessed by it its exit.

The heat possessed by the air at its entrance into the cylinder is,
$$wc^1\tau_0;$$
The internal heat at its exit is,
$$wc^1\tau_1.$$
Hence the work of compression is,
$$Jwc^1\tau_1 - Jwc^1\tau_0 = Jwc^1(\tau_1 - \tau_0),$$
and the net work is,
$$W_1 = Jc^1w(\tau_1 - \tau_0) - p_0v_0 + p_1v_1.$$
Substituting for p_0v_0 and p_1v_1 their values from eq. (34a) we have,
$$W_1 = Jwc(\tau_1 - \tau_0) \qquad (35)$$
an equation perfectly general for dry atmospheric air.

III.

WORK OBTAINABLE FROM THE COMPRESSED AIR.

If, by any process, we cause a weight of air w to pass from one reservoir, in which there is a constant pressure p_0, into another reservoir, in which there is a constant pressure p_1, and thereby consume an amount of work W_1, the same weight of air w (supposing the air to remain in the same physical conditions) will restore the amount of consumed work W_1 in passing back from the second reservoir into the first. These are the conditions of a perfect thermodynamic engine.

The work theoretically obtainable from compressed air is therefore, eq. (35),
$$W_1 = Jwc(\tau_1 - \tau_0) = W_2,$$
an equation which shows how important it is to take into account the initial and final temperature of the air.

IV.

THE THEORY OF COMPRESSION.

1. *The Work necessary, and the Volume of the Compressing-Cylinder.*—Neglecting all dead spaces and resistances, we can easily calculate, by the aid of our formulas and of Table I, the work necessary to compress to a pressure p_1 a weight of air

w, taken at a pressure p_0 and a temperature τ_0, as well as the volume to be given to the cylinder of the compressor to compress a given weight of air w per second, the time T being given in seconds.

Our formulas are:

$$W_1 = Jwc(\tau_1 - \tau_0) = Jwc\tau_0\left\{\frac{\tau_1}{\tau_0} - 1\right\}, \quad (35a)$$

when a final temperature τ_1, which is not to be exceeded, is assumed, the value of $\frac{\tau_1}{\tau_0}$ being obtained as a function of $\frac{p_1}{p_0}$ from Table 1, or from an adiabatic curve.

$$W_1 = Jwc\left\{\left(\frac{p_1}{p_0}\right)^{\frac{\gamma-1}{\gamma}} - 1\right\}, \quad (35b)$$

when a pressure p_1, to which we wish to attain, is assumed.

$$W_1 = p_0 v_0 \frac{\gamma}{\gamma-1}\left\{\frac{\tau_1}{\tau_0} - 1\right\}, \quad (35c)$$

an equation employed when we wish to find W_1 as a function of the volume v_0 of the air instead of as a function of its weight. This equation is obtained by substituting in eq. (35a.) the value of τ_0 from eq. (34a.), and γ for $\dfrac{c}{c'}$.

From eq. (34a.) we have,

$$V_1 = Rw\frac{\tau^0}{p_0} \times T, \qquad (36)$$

an equation for the volume of the cylinder which compresses per second a weight of air w, when the time, T, required per single stroke of the compressor (or per double stroke when the compressor is single-acting), is given in seconds.

2. *The Final Temperature of the Compressed Air.*—This is found by looking in Table I. for the values of $\dfrac{\tau}{\tau_0}$ opposite the different values of $\dfrac{p}{p_0}$. Supposing the initial temperature $\tau_0 = 293° = 20°C.$, we find for the different values of $\dfrac{p}{p_0}$ the values of τ_1 in degrees of absolute temperature and degrees C., as follows:

Table II.

$\frac{P_1}{P_0}$	τ_1	Final Temperature in Degrees C.
2	358.2	85.2
3	402.9	129.9
4	437.9	164.9
5	467.2	194.2
6	492.6	219.6
7	515.1	242.1
8	535.4	262.4
9	554.1	281.1
10	571.3	298.3
11	587.2	314.2
12	602.2	329.2
13	616.4	343.4
14	629.8	356.8
15	642.5	369.5

V.

THE THEORY OF TRANSMISSION.

1. *Loss of Pressure due to Transmission.*—The loss in pressure which results from carrying compressed air from one point to another point distant from the first, is due,

1.° To the friction between the air and the conveying pipes;

2.° To sudden contractions in the pipes;

3.° To sharp turns and elbows.

From experiments made at the Mont Cenis Tunnel, the loss of pressure from friction in pipes was formulated thus:—

$$\Delta p = .00936 \frac{u^2 l}{d}, \qquad (37)$$

where u = the velocity of the air per second,

l = length of the pipes,

d = diameter " "

Hence *the loss of pressure varies, directly as the length of pipe; directly as the square of the velocity of the air in the pipe; inversely as the diameter of the pipe.*

If w be the weight of air required by the working-cylinder per second, $3.1416 \frac{d^2}{4} u$ being the volume of air passing through the pipe per second, and p_1 and τ_1 being the pressure and absolute temperature respectively of the air in the reservoir, we have, from eq. (34a)

$$\frac{3.1416\frac{d^2}{4}up_1}{\tau_1} = Jw(c-c');$$

Solving with respect to u and substituting in (37), we have,

$$\Delta p = 13.88 \frac{w^2 \tau_1^2 l}{p_1^2 d^5}$$

when Joule's equivalent is taken in French units; when taken in British units (772 foot-pounds per British thermal unit), we have,

$$\Delta p = 43.055 \frac{\tau_1^2}{d^5} \frac{w^2}{p_1^2} l \qquad (38)$$

which expresses the loss of pressure due to friction in the pipes as a function of the weight of air supplied per second, of the temperature and pressure of the air in the reservoir, and of the length and diameter of the pipe.

2. *Difference of Level.*—The difference of level which exists between the reservoir and the compressor and the compressed-air engine (as when the latter is at the bottom of a mine) compensates in part, at least, for the loss of pressure due

to the friction in the supply-pipes. The gain in pressure due to this difference of level is readily calculated by means of the ordinary barometric formulae. (See Wood's Elementary Mechanics, p. 327).

VI.

THE THEORY OF COMPLETE-EXPANSIVE WORKING.

1. *Notation.*—Let θ_0 = the absolute temperature of the compressed air when it enters the working cylinder;

θ_1 = the absolute temperature of the air after expansion;

Φ_0 = the pressure of the compressed air on entering the working-cylinder;

Φ_1 = the pressure at the end of expansion.

2. *Work theoretically obtainable.*—This is given in Section III, and is:

$$W_2 = Jwc(\theta_0 - \theta_1) = Jwc\theta_0 \left\{ 1 - \frac{\theta_1}{\theta_0} \right\}$$

$$= Jwc\theta_0 \left\{ 1 - \left(\frac{\Phi_1}{\Phi_0}\right)^{\frac{r-1}{r}} \right\}, \quad (39)$$

$\dfrac{\theta_1}{\theta_0}$ being obtained from the formula for the final temperature.

3. *Final Temperature.*—This is given by eq. (34d) and is :

$$\frac{\theta_1}{\theta_0} = \left\{ \frac{\Phi_1}{\Phi_0} \right\}^{\frac{\gamma-1}{\gamma}}$$

it can be calculated directly by the use of Table I when we know $\dfrac{\Phi_1}{\Phi_0}$, the ratio of the final to the initial temperature.

4. *Volume of the Working-Cylinder.*—The volume of the working-cylinder, being the same as the final volume of the air after expansion is, from eq. (34a),

$$V_2 = Jw \frac{\theta_1}{\Phi_1} (c-c') \, T \qquad (40.)$$

where $w=$ the weight of air furnished per second and $T=$ the time in seconds of one stroke.

5. *Weight of Air required per Second.* This is determined by the work which is to be done by the compressed-air engine per second. Letting k be a certain coefficient embracing resistances of all kinds, we have, Section III,

$$w = \frac{W_2}{k\, Jc(\theta_0 - \theta_1)} \qquad (41)$$

Substituting this value of w in eq. (36) we have,

$$V_1 = \frac{R}{Jc} \times \frac{W_2 T}{k(\theta_0 - \theta_1)} \times \frac{\tau_0}{p_0} = \frac{\gamma-1}{\gamma} \times \frac{W_2 T}{k(\theta_0 - \theta_1)} \times \frac{\tau_0}{p_0}, \qquad (42)$$

the volume of the compressor in order to supply the given amount of air.

6. *Cold resulting from Expansion.*— While in the compressor there is a great development of heat from the compression of air, in the working-cylinder there is a great fall of temperature due to its expansion. The final temperature θ_1 is calculated from the formula of Sec. VI, 3.

The valves of $\frac{\theta_1}{\theta_0}$, corresponding to $\frac{\Phi_1}{\Phi_0}$, and the reciprocals, are found from Table I. The following table is from M. Mallard. The initial absolute temperature is assumed $\theta_0 = 293°$, that is, $20°$ C.

Table III.

$\dfrac{\Phi_1}{\Phi_0}$	Final Temperature.	
	Absolute θ_1.	Degrees C.
2	239.6	— 33.4
3	213.0	— 60.0
4	196.0	— 77.0
5	183.7	— 89.3
6	174.2	— 98.8
7	166.6	—106.4
8	160.3	—112.7
9	154.9	—118.1
10	150.1	—122.9
11	146.1	—126.9
12	142.5	—130.5
13	139.2	—133.8
14	136.3	—136.7
15	133.6	—139.4

This table shows what very low temperatures are reached when we work full expansion with air at a high pressure. Ice is formed from the water-vapor present in the air, and seriously interferes with the action of the working engine.

VII.

THE THEORY OF FULL PRESSURE WORKING.

1. *Work obtainable.*—This is, in the present case, expressed by the equation,

$$W_2 = V_2(\Phi_0 - \Phi_1). \qquad (43)$$

Placing in this equation the value of V_2 from eq. (40) we have,

$$W_2 = Jw(c-c')\theta_0 \left\{ 1 - \frac{\Phi_1}{\Phi_0} \right\}. \qquad (44)$$

The general expression for the work restored has been given by eq. (39), where θ_1 is the temperature of the air after it has been exhausted and has assumed the pressure of the atmosphere Φ_1.

2. *Final Temperature.*—Placing eqs. (44) and (39) equal to each other,

$$\frac{c-c'}{c}\left(1 - \frac{\Phi_1}{\gamma_0}\right) = \left(1 - \frac{\theta_1}{\theta_0}\right)$$

or, $\dfrac{\theta_1}{\theta_0} = \dfrac{1}{\gamma} + \dfrac{\gamma-1}{\gamma}\dfrac{\Phi_1}{\Phi_0} = .7102 + .29\dfrac{\Phi_1}{\Phi_0}$ (45)

3. *Weight of Air necessary per Second.*—This is given by eq. (41).

4. *Volume of Cylinder.*—Substituting w, eq. (41), in eq. (34a), we have,

$$V_1 = \frac{c-c'}{c} \times \frac{W_2 T}{\Phi_0 \left\{1 - \frac{\theta_1}{\theta_0}\right\} k} \qquad (46)$$

VIII.

THEORY OF INCOMPLETE EXPANSIVE WORKING.

1. *Work attainable.*—This is given by eq. (39).

2. *Final Temperature.*—We have, eq. (34d),

$$\frac{\theta_1'}{\theta_0} = \left\{\frac{\Phi'}{\Phi_0}\right\}^{\frac{\gamma-1}{\gamma}},$$

from which we get θ_1' (the temperature at the end of the stroke). θ_1 is then found from the equation,

$$\frac{\theta_1}{\theta_1'} = \frac{1}{\gamma} + \frac{\gamma-1}{\gamma} \frac{\Phi_1}{\Phi_0}.$$

3. *The weight of air used.*—This is given by eq. (41.)

4. *Volume of the Cylinder.*—Eq. (34a), written to satisfy our conditions, becomes:

$$V_2 = J(c-c')wT\frac{\theta_1'}{\Phi_1''},$$

or, substituting the value of w from eq. (41),

$$V_2 = \frac{\gamma-1}{\gamma} \frac{W_2 T}{\Phi_1' \frac{\theta_0}{\theta_1'} \left\{1 - \frac{\theta_1}{\theta_0}\right\}^\kappa}. \qquad (47)$$

IX.

GRAPHICAL REPRESENTATION FOR THE ACTION OF COMPRESSED AIR.

Let abscissas, in diagram on next page,* be volumes, and ordinates pressures; taking O for the origin. Through B ($p_0 v_0$) construct an adiabatic curve from its equation, (eq. 27).

"The *intrinsic energy* of a fluid is the energy which it is capable of exerting against a piston in changing from a given state as to temperature and volume, to a total privation of heat and indefinite expansion." The intrinsic of 1 lb. of air at p_0 and v_0, will be represented by the area included between the axis of

---

* For which we are indebted to Professor Frazier.

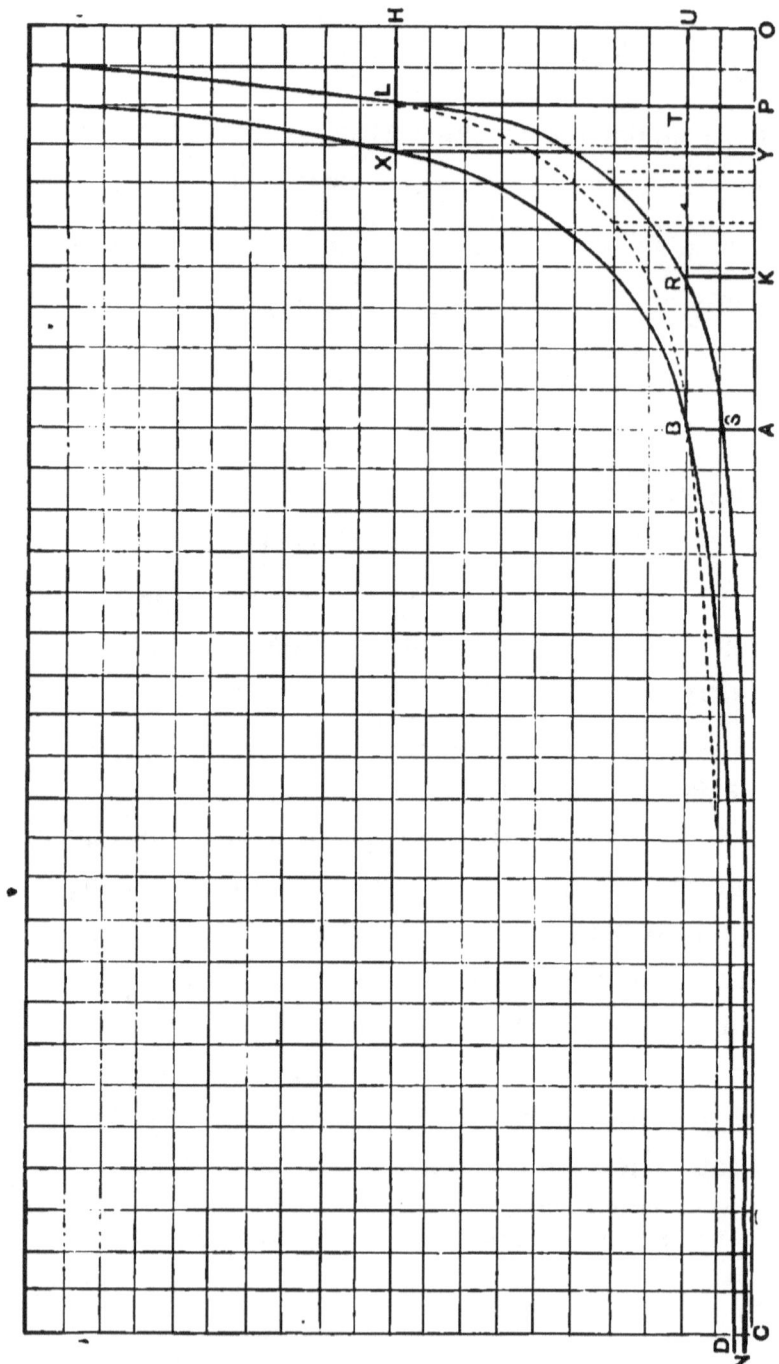

88

abscissas, the ordinate $AB = p_0$ (at a distance from the origin $OA = v_0$), and the portion of the adiabatic curve extending indefinitely from B until it becomes tangent to the axis of abscissas when $x = \infty$. The algebraic expression for this area (found by integrating eq. (27 a) between the limits ∞ and v_0 is,

$$I = \frac{p_0 v_0}{\gamma - 1}. \qquad (48a)$$

$p_0 =$ mean pressure of atmosphere in lbs. per square foot $= 2116.3$;

$v_0 =$ volume in cubic feet of 1 lb. of air at pressure p_0 and temperature τ_0 $= 12.387$;

$\tau_0 = 493.°2$ corresponding to $32°$ F;

$\gamma = 1.408$; hence

$$I = \frac{p_0 v_0}{\gamma - 1} = 64250 \text{ foot-pounds};$$

that is, one pound of air, at mean barometer pressure and $32°$F, possesses an intrinsic energy of 64250 foot-pounds; and *it is upon this store of energy that we draw, when, after abstracting in the form of heat all the work we had ex-*

pended in compressing the air, we yet cause it to perform work by expansion.

Through B construct an isothermal curve from its equation (eq. 1). At a point (as L) chosen arbitrarily upon this curve to correspond to a desired pressure we can construct another adiabatic curve LRN. Then will the relations exist, expressed as follows, and given by Prof. Frazier:

Area ABDC prolonged indefinitely = intrinsic energy possessed by the air before compression = I.

Area ABLPA = the work performed in compressing the air.

Area DBLRN prolonged indefinitely = ABLPA = energy in the form of heat abstracted by the cooling water; consequently, BSND prolonged indefinitely = ASLPA.

Area CKRN prolonged indefinitely = intrinsic energy of the air after expansion.

Area KRLPK = work performed by the air in its expansion.

Area ABRKA = work performed by the air after it leaves the working-cylinder.

Area DBRSN prolonged indefinitely = ABRLPA = the heat absorbed by the air after leaving the working-cylinder.

For isothermal compression, we have,

Area ABLHOA = total work performed in the compressing-cylinder.

Area ABLPA = work performed in the compression of the air.

Area PLHOP = work performed in the expulsion of the air from the compressor.

Area ABUOA = work performed by the atmosphere.

Area UBLHU = ABLPA = the work performed by the motor.

Area UTLHU = useful work performed by the air (full pressure).

Area UBLHU − UTLHU = TLBT = amount of work lost.

For adiabatic compression we have :

Area ABXYA = work performed in the compression of the air.

Area YXHOY = work performed in the expulsion of the air from the compressor.

Area ABUOA = work performed by the atmosphere.

Area BXHUB = work performed by the motor.

Area TLHUT = useful work performed by the air (full pressure).

Area BXLTB = BXHUB − TLHUT = amount of work lost.

When the air is allowed to expand fully (to its original pressure p_v),

Area RTLR = useful work of expansion.

Area UHLRU = total useful work (= UTLHU + RTLR).

Area BXLRB = BXHUB − UHLRU = amount of work lost where air is cooled after leaving the compressor.

Area BLRB = UBLHU − UHLRU = amount of work lost where air is cooled completely in compressor.

The area BLRB represents, then, the excess of work performed on the air above that performed by it, or the amount of work permanently transformed into heat. It is, therefore, not possible, even by preventing any rise of temperature during compression and allowing the air to expand to its full extent, to obtain from the compressed air as much work as was expended in the compression. We can obtain from compressed air all the work expended upon it, only by causing it to reproduce exactly during its expansion the changes of condition it underwent during compression. This may theoretically be accomplished in three ways.

1. By allowing the compressed air to become heated during compression, and preventing all transmission of heat until it leaves the working cylinder. It will be compressed and will expand in this case following the curve BX.

2. By cooling the air during compression and heating it during its expansion, in such a manner that its temperature

shall remain constant during both operations. The air will be compressed and will expand in this case, following the curve BL. The heat abstracted during compression will equal that supplied during expansion.

3. By cooling the air *before* its compression to such a degree that after it is compressed it will have the temperature of the media surrounding the working cylinder. The air will be compressed and will expand in this case, following the curve RL.

CHAPTER VI.

Efficiency Theoretically Attainable.

I.

EFFICIENCY OF THE AIR-COMPRESSOR AND COMPRESSED-AIR ENGINE AS A SYSTEM.

$$\frac{\text{Work performed on the air}}{\text{Work performed by the air}} =$$

the efficiency $= E$;

hence,

$$E = \frac{W_2}{W_1} = \frac{Jc(\theta_0 - \theta_1)w}{Jc(\tau_1 - \tau_0)w} = \frac{\theta_1 \left\{ \frac{\theta_0}{\theta_1} - 1 \right\}}{\tau_0 \left\{ \frac{\tau_1}{\tau_0} - 1 \right\}}$$

$$= \frac{\theta_1}{\tau_0} \times \frac{\left\{ \frac{\Phi_0}{\Phi_1} \right\}^{\frac{\gamma-1}{\gamma}} - 1}{\left\{ \frac{p_1}{p_0} \right\}^{\frac{\gamma-1}{\gamma}} - 1} \qquad (48.)$$

In practice, $\frac{\Phi_0}{\Phi_1}$ and $\frac{p_0}{p_1}$ differ very little in value, their difference being due to the loss of pressure from the friction between the air and the supply-pipe, a loss which is very small if the pipes are of sufficient diameter.

Hence we may write,

$$E = \frac{\theta_1}{\tau_0}, \qquad (48a)$$

that is to say, when compressed air is made to expand completely, and when the ratio of its pressure to the pressure of the surrounding atmosphere is the same when the air leaves the compressor as when it enters the cylinder of the compressed-air engine, *the efficiency of the system is the ratio of the temperature of the compressed air when it leaves the compressed-air-engine cylinder to the temperature of the air at its entrance into the compressor.*

This law is independent of any heat lost by the air in passing from one cylinder to the other.

Since we have just admitted that,

$$\frac{\Phi_1}{\Phi_0} = \frac{p_0}{p_1}$$

we have,

$$\frac{\theta_1}{\theta_0} = \left\{\frac{\Phi_1}{\Phi_0}\right\}^{\frac{\gamma-1}{\gamma}} = \left\{\frac{p_0}{p_1}\right\}^{\frac{\gamma-1}{\gamma}} = \frac{\tau_0}{\tau_1};$$

hence,

$$E = \frac{W_2}{W_1} = \frac{\theta_1}{\tau_0} = \frac{\theta_0}{\tau_1}, \qquad (48b)$$

showing that *the loss of work is proportional to the loss of heat undergone by the compressed air in its passage from the compressor to the working-cylinder.*

The efficiency will be a maximum when $\tau_1 = \theta_0$; that is, when the loss of heat is nothing. Of course, this condition cannot be realized. Generally the compressed air reaches the working cylinder with a temperature equal to that of the surrounding atmosphere. The temperature θ_0 is therefore given, and the efficiency can only be increased by diminishing τ_1.

The following table is calculated from (eq. 48b) for different values of $\dfrac{p_1}{p_0}$, the temperature of the compressed air at entering the working cylinder being taken $\theta_0 = 293°$, that is, 20° C.

TABLE IV.

$\dfrac{p_1}{p_0}$	E.	$\dfrac{p_1}{p_0}$	E.
2	.82	9	.53
3	.72	10	.51
4	.67	11	.50
5	.63	12	.49
6	.60	13	.48
7	.57	14	.47
8	.55	15	.46

The table shows that when the pressure has reached four atmospheres, even a considerable increase of it does not much effect the efficiency.

II.

MAXIMUM EFFICIENCY CALCULATED FROM THE INDICATED WORK.

Let $p=$ the pressure of the compressed air,

Let $p_0=$ the pressure of the atmosphere,

v and $v_0=$ the corresponding volumes; also let $p=np_0$; then $v=nv_0$.

The work spent upon the air to compress it, is, (eq. 26a),

$$W_1 = p_0 v \text{ nap. log. } \frac{v}{v_0} = p_0 v \times 2.303 \text{ com. log. } n$$

The work performed by the air is:

$$W_2 = (p-p_0)v_0,$$

and as $pv_0 = v_0 v$ and $v = nv_0$, we have

$$W_2 = p_0 v \left\{ 1 - \frac{1}{n} \right\};$$

hence,

$$E = \frac{W_2}{W_1} = \frac{p_0 v \left\{1 - \frac{1}{n}\right\}}{p_0 v \times 2.303 \text{ com. log. } n} = \frac{\left\{1 - \frac{1}{n}\right\}}{2.303 \text{ com. log. } n} \quad (49.)$$

Substituting different values of n in this formula we get the corresponding values of E.

III.

THE EFFICIENCY OF COMPLETE EXPANSION AND OF FULL PRESSURE COMPARED.

To show the comparative merits and demerits of full pressure and complete expansion in the use of compressed air, we present a table prepared by M. Mallard:

Table V.

$\frac{\Phi_6}{\Phi_1}$	Final temperature. Degrees C. Complete expansion.	Theoretical efficiency with complete expansion.	Final temperature. Degrees C. Full pressure.	Theoretical efficiency with full pressure.	Ratio of efficiency at full pressure to efficiency at complete expansion.
2	—33.4	.855	—22.4	.82	.95
3	—60.0	.806	—36.9	.72	.90
4	—77.0	.782	—43.2	.67	.86
5	—89.0	.768	—48.0	.63	.82
6	—98.0	.758	—51.0	.60	.79
7	—106.0	.751	—53.0	.57	.74
8	—112.7	.746	—54.5	.55	.73
9	—118.1	.742	—55.6	.53	.71
10	—122.9	.739	—56.5	.51	.69
11	—126.9	.736	—57.4	.50	.68
12	—130.5	.734	—58.0	.49	.66
13	—133.8	.732	—58.6	.48	.65
14	—136.7	.730	—59.2	.47	.64
15	—139.4	.729	—59.5	.46	.63

The initial temperature is assumed at 20°C.

The table shows that by working non-expansively we avoid very low temperatures of exhaust; but this is of little practical importance when we take into account the low efficiency of full pressure, as compared with complete expansive working. Also when working at full pressure, the higher the working pressure the lower the efficiency.

CHAPTER VII.

The Effects of Moisture, of the Injection of Water, and of the Conduction of Heat.

I.

GENERAL STATEMENT.

In dealing with compressed air we must always keep in view the very important consideration of the *initial* and *final temperature* of the air.

There are two principal causes tending

to vary the amount of heat present in the compressor or absorbed in the working-cylinder:—

1. The water or water-vapor of which atmospheric air always contains more or less, and which is purposely introduced into the cylinder of the so-called wet-compressors.

2. The conduction of heat by the cylinders, supply-pipes, reservoirs, &c.

II.

THE EFFECTS OF MOISTURE.

Atmospheric air always contains more or less moisture. We wish to consider the effects of this moisture upon the air undergoing compression or expansion. The injection of water into the cylinders and its cooling or heating effects are left out of the question altogether, as they will receive attention further on.

In all conditions of temperature and pressure practically realizable, a mixture of air and saturated water-vapor will remain saturated when the mixture expands against a resistance, a certain quantity of water being thereby condensed; on the contrary, compression superheats the vapor, which then becomes non-saturated, and non-saturated vapors follow the laws of permanent gases.

1. *Influence of water-vapor upon the work spent on the air and upon that performed by it.*—The presence of moisture in the air has been found to be favorable both in the compressor-cylinder and in the working cylinder. In both cases, however, the gain in work spent or performed is so slight that it can be entirely neglected, and the formulas already established for dry air become applicable with a sufficiently close approximation. In the case of compression, the vapor is superheated and therefore comports itself very much like the air itself; while in the working-cylinder, the increase of work performed, when the initial temperature of the compressed air does not exceed 30° C., is very small; and, as the temperature at

which compressed air is used, is rarely higher than 20° C., the influence of the water-vapor can be safely neglected.

2. *Influence of the moisture of the air upon the Final Temperature.*—The presence of the moisture in the atmospheric air introduced into the compressor tends to lessen the heat of compression; this effect, however, is very slight, and, in a practical point of view, is not worth considering.

When compressed air is completely expanded in a working-cylinder, the presence of moisture in it tends to lessen the cold produced. M. Mallard has found what the initial pressure would be for certain initial temperatures, so that the final temperature should not fall below 0° C. He has found this for both dry and saturated air, and his results are tabulated as follows:—

TABLE VI.

Final temperature. Degrees C.	Initial temperature. Degrees C.	Value of $\frac{\Phi_0}{\Phi_1}$ with the air.	
		Saturated with water-vapor.	Dry.
0°	20°	1.50	1.276
0°	30°	1.89	1.432
0°	40°	2.39	1.602
0°	50°	3.06	1.780

This table shows that, if compressed air at 50°C and at a pressure of three atmospheres be introduced into a working-cylinder, this air, if saturated with aqueous vapor, can be completely expanded without falling to a temperature below 0°C; and that this air, if dry, dare not exceed an initial pressure of 1.78 atmospheres if its final temperature is not to fall below 0°C.

3. *Volume of the Cylinders.*—This is calculated as for dry air, since the effect of the moisture is too slight to be taken into account.

III.

THE INJECTION OF WATER.

1. *The Effect of Introducing Water into the Compressor-Cylinder.*—It is of great advantage in practice to introduce cold water into the compressor. It carries away the heat of compression to a very great extent. It acts as a lubricant, and, by cooling the cylinder, it prevents the destruction of any organic material, such as packing, valves, &c., that may be employed upon it.

If in addition to the atmospheric moisture present in the air at its entrance into the compressor, water be introduced in quantities just sufficient to keep the air saturated with water-vapor during the compression, the work spent upon the air and the final temperature at the end of compression will both be less than if the air had not been kept saturated while being compressed. It is unnecessary to calculate the amount of work saved or the extent to which the temperature is reduced by the presence of this saturated water-vapor; for if water is at all to be introduced into the compressor, it may as well be thrown in in larger quantities, that is, in quantities sufficient to absorb and carry off the greater part of the heat of compression.

The effects of the heated air in the compressor is a great cause of loss of motive power, and it is very desirable to cool the air during its compression.

The final temperatures for different pressures have already been given in Table II. We repeat them here in connection with the quantities of work spent when the compression follows Boyle's law and when it is effected without any removal of heat.

Table VII.

Tension in atmospheres.	Compression with temperature constant.		Compression with increase of temperature.			Loss of work due to the heat of compression. kilogrammeters.	Fraction of the total work required for compression, which is converted into heat.
	Volume in cubic meters.	Work in kilogrammeters.	Temperature in Degrees C.	Volume in cubic meters.	Work in kilogrammeters.		
1	1.00		20°	1.			
2	.50	7,199	85°.5	.612	7,932	733	.092
3	.333	11,356	130°.4	.459	13,360	2004	.150
4	.250	14,260	165°.6	.374	17,737	3477	.196
5	.200	16,580	195°.3	.320	21,209	4629	.213
6	.167	18,475	220°.5	.281	24,310	5835	.240
7	.143	20,038	243°.2	.252	27,048	7040	.260
8	.125	21,422	263°.6	.229	29,518	8096	.274

The Quantity of Water to be Injected.
—We have found eq. (26), that the quantity of heat developed by compression is given by the formula,

$$Q = \frac{R\tau_0}{J} \text{ nap. log.} \left\{ \frac{v_0}{v_1} \right\},$$

where τ_0 is the absolute final temperature $=273°+40°=313°$. From this formula the quantity of heat, Q, is calculated for different pressures. We then find the weight of water, which, if introduced at 20°C and removed when it has taken up enough heat to raise its temperature to 40°C, would absorb this quantity of heat Q. Under these conditions we find that each kilogramme of water will absorb 20 calories. Dividing Q by 20 we get the weight of water to be introduced in kilogrammes. In this way the following table was prepared:

Table VIII.

Absolute pressure to which the air is compressed.	Heat developed by compression and to be carried off by the injected water.	Weight of water at 20° C. to be injected into the compressor per kilogramme of air compressed in order to keep the final temperature from rising above 40° C.
atmospheres.	calories.	kilogrammes.
2	14.695	.734
3	23.284	1.164
4	29.392	1.469
5	34.120	1.701
6	37.979	1 891
7	41.264	2.063
8	44.087	2.204
9	46.589	2.329
10	48.816	2.440
11	50.849	2.542
12	52.694	2.634
13	54.391	2 719
14	55.962	2.798
15	57.425	2.871

2. *The Injection of Hot Water into the Cylinder of the Compressed-Air Engine.*—In the production of compressed air, the great cause of loss of motive power, as we have seen, is the development of heat. Analogous to this is the

loss which occurs in the *use* of compressed air. Great cold is produced by expansive working, and this has long forbidden its adoption. The injection of hot water into the working-cylinder, has now made it possible to attain the desirable result of working expansively.

The Quantity of Hot Water to be Introduced.—The quantity of heat, Q, to be supplied to keep the temperature of the expanding air constant is found from eq.(26), to be,

$$Q = \frac{R\tau_0}{J} \text{ nap. log.} \left\{ \frac{v_1}{v_0} \right\}.$$

The expansion being supposed to follow Boyle's law, we have,

$$p_1 v_1 = p_0 v_0, \text{ or } \frac{v_1}{v_0} = \frac{p_0}{p_1}$$

Hence we have,

$$Q = \frac{R\tau_0}{J} \text{ nap. log.} \left\{ \frac{p_0}{p_1} \right\}.$$

$p_1 = 1$ in this case since the air is expanded down to atmospheric pressure. From this formula the weight of water to be injected is calculated as in table. The results are given in the following:

Table IX.

Absolute pressure at which the compressed air is introduced into the working cylinder.	Quantity of heat to be supplied to keep the temperature of the air from falling below 0° C. during its expansion down to atmospheric pressure.	Weight of water to be injected into the working cylinder per kilogramme of compressed air introduced to keep the final temperature from falling below 0° C.		
		The temperature of the water introduced being		
		20° C.	50° C.	100° C.
2	13.280	.134	.103	.074
3	21.030	.212	.163	.117
4	26.550	.262	.206	.148
5	30.828	.311	.240	.178
6	34.334	.346	.266	.192
7	37.285	.376	.289	.208
8	39.833	.402	.309	.223
9	42.094	.425	.326	.235
10	44.106	.445	.342	.247
11	45.945	.464	.356	.256
12	47.612	.480	.369	.266
13	49.145	.496	.381	.274
14	50.562	.510	.392	.282
15	51.885	.524	.402	.290

The quantities of water here given are the minima values, since the latent heat which is released by the water in freezing has not been taken into account. Hence to avoid the formation of ice we must add a slight excess of hot water.

3. *The Effect of the Conduction of Heat by the Cylinders, Pipes, &c.*—Since the temperature of the compressed air when used is most always that of the surrounding atmosphere, the result of the conduction of heat by the containing vessels is the dissipation of the total heat of compression. The mechanical equivalent of this heat is, of course, lost work, and, as it is most economical to get rid of this heat during compression, conduction and radiation from the compressor is an advantage. Since, in working expansively, there is a tendency for the cylinder to become colder than surrounding bodies, the conduction and radiation of heat is here too, if anything, an advantage.

In all our formulas and results hitherto established, the cylinders have been supposed non-conducting; and the investigations of M. Mallard have shown that this hypothesis is justified. For the heat leaving the compressor by conduction and radiation is in part compensated for by that developed by the friction of the piston; and the heat conducted through the working cylinder is very small relatively to that converted into work. Hence, any passage of heat by conduction of the cylinders belongs to those secondary quantities which are always omitted in the general theory of motors, except so far as allowed for by proper coefficients.

CHAPTER VIII.

AMERICAN AND EUROPEAN AIR-COMPRESSORS.

I.

PUMP COMPRESSORS.

Pump or plunger compressors are generally in high repute in Germany and Austria, especially in mines, and they seem to give very satisfactory results. In the United States they never have been used to any considerable extent and are now not at all used.

It must be said to the prejudice of these compressors, that, in consequence of the large mass of water to be pushed back and forth by the plunger, a large per-centum of power is wasted in overcoming inertia; that high piston speeds are, in consequence of the violent shocks which result, utterly impossible; that they are very heavy and hence require expensive foundations; that when the prime mover is run at a high speed, a more or less cumbrous, expensive, and wasteful machinery of transmission is necessary; that their use is limited, pressures of 5 or 6 atmospheres being their utmost capability, on account of the large quantity of cooling water taken up by the air at even moderately high tensions; that a large amount of cooling-water is required to produce a comparatively small effect in the abstraction of heat.

On the other hand, it must be admitted that these compressors are liable to very few repairs, that they are simple in construction and that "dead spaces" are avoided.

The hydraulic or ram compressors first used by Sommeiller at the Mt. Cenis Tunnel have become obsolete.

II.

SINGLE-ACTING WET COMPRESSORS.

The air compressors now used in the United States are either, *Dry Compressors* in which the cooling is effected by flooding the external of the cylinder, and sometimes also the piston and piston-head with water; *Wet Compressors*, by the injection of water into the cylinder-space, as well as by external flooding; *compressors* with *no* cooling arrangement are seldom used, and only in temporary and cheap plants.

Compressors with a partial injection of water have been used to very good effect in the United States. Most of these are single-acting, and are represented by the machines of Burleigh, of Fitchburgh, Mass. The cooling is very efficient and hence the useful effect is considerably increased. They are very durable and not liable to get out of repair, as is shown by the record of Burleigh's machines, which have stood the test of years of steady work.

The use of single-acting compressors renders it necessary that, in all cases where anything like a uniform supply of air is needed, to have two compressor-cylinders. These cannot be driven directly from the piston-rod of the driving engine, but necessitate an indirectly coupled-connection of some sort. All this makes single-acting compressors somewhat cumbrous and expensive.

As built to-day, the evils of dead spaces, and of jars and shocks resulting from water in the cylinder, have not been duly considered. There are also a few cases when the sectional area of the inlet-valves is insufficient; and in general those parts which are most liable to get out of repair are most difficult of access.

We are inclined to think that the claim of the Burleigh Co., that their compressor is the most efficient, economical, and durable of any built in this country, cannot be far from the truth.

III.

DOUBLE AND DIRECT-ACTING COMPRESSORS.

Up to within several years ago, single-acting compressors have been used almost exclusively. Now the double and direct-acting compressor seems to be superseding it. This is now the leading type of American compressor, although hitherto it has given at least no better results than the best single-acting machine.

Superiority in the double-acting compressor is found in its simplicity. The piston of the engine drives the compressor by a direct connection. All wasteful and cumbrous machinery of transmission is at once unnecessary and high piston-speeds are possible; in the United States from five to seven feet.

Most American double and direct-acting compressors are of the dry kind. These have the advantage that the air is delivered without having any water mechanically mixed with it. Hence very much ice cannot be formed when the air is worked expansively. Higher rates of expansion are possible than with air from a wet compressor.

One of the very best American double and direct-acting dry compressors is the "National," built by Allison & Brannan, Port Carbon, Pa., (Office, 95 Liberty St., N. Y.). Steam cylinders of the medium-sized duplex machine are $12'' \times 42''$, and the air cylinders $15'' \times 42''$. The air pistons work to within one sixteenth of an inch of the cylinder heads. The water circulation for cooling passes spirally around the air cylinder from the center to each end. The engine will compress air to the same pressure as that of the steam used. The amount of free air compressed at a piston speed of 350 feet is about 1000 cubic feet per minute. A greater pressure of air than the pressure of steam used is obtained by increasing the size of the steam cylinder, or decreasing that of the air cylinder.

The best double and direct acting compressor of the wet kind is undoubtedly that of Dubois-François, built in Seraing, Belgium, and exhibited at the Centennial Exposition, in 1876.

Dry compressors, although the cheapest as regards first cost, are not the most economical in working. But where air is to be carried through pipes exposed to great cold they are the only alternative.

IV.

DESIGN AND CONSTRUCTION.

The efforts of builders and engineers should be directed to the attaining of a higher efficiency, and they should not, as is now often the case, sacrifice the latter to cheapness and small dimensions. To attain such desirable efficiency the heat of compression must be more effectually abstracted. This must be done by a more ingenious and rapid circulation of water around the cylinder, and injection of water in the form of spray into the cylinder. But the injection of water in some efficient and practical manner, which is so essential to the reaching the highest efficiency, introduces the great disadvantage of having to work with wet air. Hence we see how important would be an invention of means or apparatus for separating the water from the air when direct intercontact has been had to keep down the temperature. We must also remember the important physical fact that water absorbs very considerable volumes of air—volumes dependent upon the pressure of the air and the amount of surface of water exposed to the fluid contact, time being also an important factor.

Clearance must be reduced to the smallest possible amount. It has been brought down in a few cases to 0.39 inch. A long stroke, one from 2 to to 3 times the diameter of the cylinder, is another means of avoiding loss from dead spaces, since here the air which fills the dead space is small in comparison with that actually delivered. The valves must be so placed that, between their seats and the piston-head at the end of the stroke there shall be the smallest possible clearance.

The valves themselves, to close the more rapidly, are made to have only a very small travel. (This has been made as small as .08 to .12 inch.) The valve-area must be made large enough by increasing the number of the valves. It should be amply large, generally from $\frac{1}{5}$ to $\frac{1}{10}$ of the sectional area of the cylinder. The valves should be so attached to the cylinder-head that they may be removed and repaired without taking off the latter or otherwise taking the machine apart.

Great care must be taken to have the piston head fit the cylinder accurately and closely, since, especially in dry compressors, great losses result from any looseness. The piston-heads should be made so that they can be adjusted to preserve a nice fit, as in steam engine practice. Lubrication of the cylinder in case of the dry compressor should be effected by automatic oil cups placed upon it.

It must also be borne in mind that the working pressure is that which most influences the physical conditions of working, and the suitable mode of construction. And, although the loss of work increases with the pressure, yet the *rate of variation* of the loss of work decreases as the pressure increases. As great a proportion of work is lost by increasing the pressure from two to three atmospheres as by increasing it from five to ten atmospheres.

The tendency in Germany and France, as well as here, is for the wet compressor entirely to supersede all others. But it is scarcely too much to say that the air-compressor of the future has yet to be invented.

CHAPTER IX.

EXAMPLES FROM PRACTICE.

I.

The Republic Iron Company of Marquette, Mich., have done away with the use of steam, by utilizing the power of a water-fall situated about a mile from their works. The power is transmitted by means of compressed air which drives all their machinery, and thus saves the cost of fuel.

There are four compressors, 24″ diameter and 5′ stroke, driven by two turbine Swain water-wheels $5\frac{1}{2}′$ diameter, under 16 feet head of water. As near as has been ascertained, they have about 450 horse power at the wheels. The air is carried one mile in a pipe built of boiler iron, 15″ inside diameter. About 66 per cent. of the effective power of the wheels is obtained at the mines and shops.

II.

ECONOMY PROMOTED BY THE USE OF COMPRESSED AIR.

To show the great saving of both time and money since the introduction of compressed-air machinery we will give a few figures.

It cost the Golden Star Mining Co., of Sacramento $12 to $15 per foot to run a tunnel 7×7 feet, when employing hand labor; after introducing air machinery it cost them $6 to $7 per foot; with hand labor they made a distance of two feet per day; with machine labor, a distance of six feet per day.

Another instance, among many, is that of the Sutro Tunnel Company of Nevada;

Expense by hand labor per month....................$34,000 to $50,000
Expense by machine labor per month................$14,000 to $16,000

III.

COMPRESSED-AIR MOTOR STREET CAR.

The pneumatic engine which has been

on trial by the Second Avenue Railroad Company, on the Harlem portion of their road, from the Station at Ninety-Sixth Street, to Harlem River, at One-Hundred-and-Thirtieth Street, has proved so satisfactory to the company that it has authorized the construction of five more engines.

These are to be used exclusively on the upper part of the road, where it is proposed to dispense entirely with the use of horse power, so soon as the requisite number of engines shall be procured. It was stated at the company's office, that the most sanguine expectations had been fulfilled; the new engine could be run at a trifling cost, and without the noise and smoke and smell of oil which accompany the use of steam; any rate of speed which was likely to be required could be maintained, and the engine was under as complete control of the engineer as one propelled by steam or a car drawn by horses.

The new engines are manufactured by the Pneumatic Tramway Engine Company, whose office is at No. 317 Broadway. Some time ago two Scotch engineers, Robert Hardie and J. James, invented a system of propelling cars by means of compressed air. The invention was examined by a number of practical railroad men who were visiting Scotland. Hardie and James were induced to visit this country and the company was organized. Experiments have been making for a year, resulting in improvements which now seem likely to render the invention serviceable to the public. The motive power is condensed air, contained in two reservoirs, placed one under each end of a car, which is similar in construction to those in ordinary use on street railways. The air is pumped in by a stationary engine at one hundred and twenty-seventh street, and this has been so far improved that the reservoirs in the cars now used are filled in a few minutes. These are of steel, and are tested up to a strength many times greater than their working pressure, and it is claimed that there is no danger of explosion. The machinery is simple and not liable to get out of order. The air-tanks of the experimental car are only sufficiently large to enable it to make one round trip between Harlem and Ninety-Sixth Street stations; but the cars now building will be larger and will contain reservoirs of much greater capacity; and it is claimed that there will be no difficulty in constructing them so that the round trip from Harlem river to Peck Slip can be made without replenishing.

Mr. Henry Bushnell, of New Haven, is the inventor and constructor of another new compressed air motor street car, the chief peculiarity of which is that he is able, as he says, to force air into his receivers until his gauge registers the enormous pressure of more than 3,000 pounds per square inch. His receivers are tubes, the largest of which are twenty feet long, and only eight inches in diameter, inside measurement. There are four of these, two lying side by side above the axles, and next to the wheels on either side of the car. Between them at one end are four other tubes, each six feet long and six inches in diameter, inside measurement. The material is wrought iron three-eighths of an inch thick, and are welded in. The double cylinder engine which utilizes this air in turning the wheels of the car does not differ materially from a steam engine, except that its two cylinders are only two and three-fourths inches in diameter, inside measurement. The machine built by Mr. Bushnell to compress the air consists of three steam air pumps. The first and largest is merely a feeder to the second. The air that comes from it is condensed to a pressure of about six pounds. This denser air is more worthy the prowess of the second pump, which in turn crushes it into a greatly smaller compass. The third pump gives the final pressure. The gauge on the compressing machine has registered 3,500 pounds per square inch. The plungers of the second and third pumps have no heads. They are merely rods of steel forced into vessels containing oil. As the plungers move out and in, the surface of the oil falls and rises, admitting the air through one valve and forcing it out of another. It is, therefore, necessary to have the packing of the plungers only oil tight, not air tight, under the tremendous pressure. The chamber that first receives the air from the third pump is cooled by a covering of cotton waste saturated with water. On the other hand, the expansion of the air as it is given off at each half revolution of the car engines absorbs heat, and after running the car for a short time the engine cylinders and escape pipes are whitened with frost. To remedy this Mr. Bushnell will surround the cylinders with stout metal jackets, beneath which he will force air with the aid of a small pump geared to the machinery of the car. This newly-compressed air, he says, will supply heat enough to keep the cylinders warm.

The writer rode recently on the new car as far on the Whitneyville road as Mr. Bushnell could go without interfering with the trips of the horse cars. The motion was easy, and at times about twice as rapid as that of a horse car. The new vehicle obeyed the engineer promptly in starting and stopping. The distance traveled in going and returning was a little over a mile. At the start the guage registered 1,800 pounds. At the return the pressure indicated was 1,500 pounds. When the air was allowed to escape from a turned cock the roar was frightful and was as irritating to the ear as escaping steam. In running, however, very little noise is heard from the escape-pipe, because the escaping air is made to pass through a mass of ordinary curled hair. This device Mr. Bushnell esteems one of the most important of his inventions. He has no doubt that it would prove equally efficacious in deadening the sound of escaping steam.

Friends of Mr. Bushnell claim that he could never make a receiver capable of retaining air at the high pressure he had in view. The air that was in the tubes was pumped in, he says, on the 25th of June. The gauge then showed 2,100 pounds. The pressure gradually lessened until it was 1,900. After that time a small leak was discovered. This leak was closed with a turn of the wrench, and after that not a pound was lost up to the trial, when 100 pounds was allowed to blow off to gratify the curiosity of visitors just previous to the short trip referred to.

Mr. Bushnell called attention to the small diameters of his largest tubes. He said that a pressure of 2,000 pounds per square inch would give, by calculation on the head of each tube, an aggregate pressure of fifty tons; while the two-feet heads used by the inventor of a rival compressed air motor would have to withstand an aggregate pressure of 180 tons, if a pressure of 800 pounds per square inch should be put on, as the inventor claimed was possible. The heads were necessarily the weakest parts of the tubes. A welded joint, such as his were, was usually reckoned twice as strong as a riveted one.

On a previous occasion Mr. Bushnell made a round trip on his car on the Whitneyville road, a distance of a little over four miles. The pressure was then reduced from 1,950 pounds at the start to 750 pounds on the return. A company called the United States Motor Power Company has been formed, and Mr. Bushnell is its president.

∗ *Any book in this Catalogue sent free by mail, on receipt of price.*

VALUABLE
SCIENTIFIC BOOKS,

PUBLISHED BY

D. VAN NOSTRAND,

23 Murray Street, and 27 Warren Street,

NEW YORK.

WEISBACH. A MANUAL OF THEORETICAL MECHANICS. By Julius Weisbach, Ph. D. Translated by Eckley B. Coxe, A.M., M.E. 1100 pages and 902 wood-cut illustrations. 8vo, cloth, ... $10 00

FRANCIS. LOWELL HYDRAULIC EXPERIMENTS—being a Selection from Experiments on Hydraulic Motors, on the Flow of Water over Weirs, and in open Canals of Uniform Rectangular Section, made at Lowell, Mass. By J. B. Francis, Civil Engineer. Third edition, revised and enlarged, with 23 copper-plates, beautifully engraved, and about 100 new pages of text. 4to, cloth, ... 15 00

KIRKWOOD. ON THE FILTRATION OF RIVER WATERS, for the Supply of Cities, as practised in Europe. By James P. Kirkwood. Illustrated by 30 double-plate engravings. 4to, cloth 15 00

D. VAN NOSTRAND'S PUBLICATIONS.

FANNING. A PRACTICAL TREATISE OF WATER SUPPLY ENGINEERING. Relating to the Hydrology, Hydrodynamics, and Practical Construction of Water-Works, in North America. With numerous Tables and 180 illustrations. By J. T. Fanning, C.E. 650 pages. 8vo, cloth extra, . . $6 00

WHIPPLE. AN ELEMENTARY TREATISE ON BRIDGE BUILDING. By S. Whipple, C. E. New Edition Illustrated. 8vo, cloth, 4 00

MERRILL. IRON TRUSS BRIDGES FOR RAILROADS. The Method of Calculating Strains in Trusses, with a careful comparison of the most prominent Trusses, in reference to economy in combination, etc., etc. By Bvt. Col. William E. Merrill, U. S. A., Corps of Engineers. Nine lithographed plates of illustrations. Third edition. 4to, cloth, ... 5 00

SHREVE. A TREATISE ON THE STRENGTH OF BRIDGES AND ROOFS. Comprising the determination of Algebraic formulas for Strains in Horizontal, Inclined or Rafter, Triangular, Bowstring Lenticular and other Trusses, from fixed and moving loads, with practical applications and examples, for the use of Students and Engineers By Samuel H. Shreve, A. M., Civil Engineer Second edition, 87 wood-cut illustrations. 8vo, cloth, . . . 5 00

KANSAS CITY BRIDGE. WITH AN ACCOUNT OF THE REGIMEN OF THE MISSOURI RIVER,— and a description of the Methods used for Founding in that River. By O. Chanute, Chief Engineer, and George Morison, Assistant Engineer. Illustrated with five lithographic views and twelve plates of plans. 4to, cloth, . . . 6 00

D. VAN NOSTRAND'S PUBLICATIONS.

CLARKE. DESCRIPTION OF THE IRON RAILWAY BRIDGE Across the Mississippi River at Quincy, Illinois. By Thomas Curtis Clarke, Chief Engineer. With twenty-one lithographed Plans. 4to, cloth, . . . $7 50

ROEBLING. LONG AND SHORT SPAN RAILWAY BRIDGES. By John A. Roebling, C. E. With large copperplate engravings of plans and views. Imperial folio, cloth, . 25 00

DUBOIS. THE NEW METHOD OF GRAPHICAL STATICS. By A. J. Dubois, C. E., Ph. D. 60 illustrations. 8vo, cloth, ... 2 00

EDDY. NEW CONSTRUCTIONS IN GRAPHICAL STATICS. By Prof. Henry B. Eddy, C. E Ph. D. Illustrated by ten engravings in text, and nine folding plates. 8vo, cloth, 1 50

BOW. A TREATISE ON BRACING—with its application to Bridges and other Structures of Wood or Iron. By Robert Henry Bow, C. E. 156 illustrations on stone. 8vo, cloth, 1 50

STONEY. THE THEORY OF STRAINS IN GIRDERS—and Similar Structures—with Observations on the Application of Theory to Practice, and Tables of Strength and other Properties of Materials. By Bindon B. Stoney, B. A. New and Revised Edition, with numerous illustrations. Royal 8vo, 664 pp., cloth, ... 12 50

HENRICI. SKELETON STRUCTURES, especially in their Application to the building of Steel and Iron Bridges. By Olaus Henrici. 8vo, cloth, ... 1 50

KING. LESSONS AND PRACTICAL NOTES ON STEAM. The Steam Engine, Propellers, &c., &c., for Young Engineers. By the late W. R. King, U. S. N., revised by Chief-Engineer J. W. King, U. S. Navy. 19th edition. 8vo, cloth, ... 2 00

AUCHINCLOSS. APPLICATION OF THE SLIDE VALVE and Link Motion to Stationary, Portable, Locomotive and Marine Engines. By William S Auchincloss. Designed as a hand-book for Mechanical Engineers. With 37 wood-cuts and 21 lithographic plates, with copper-plate engraving of the Travel Scale. Sixth edition. 8vo, cloth, $3 00

BURGH. MODERN MARINE ENGINEERING, applied to Paddle and Screw Propulsion. Consisting of 36 Colored Plates, 259 Practical Wood-cut Illustrations, and 403 pages of Descriptive Matter, the whole being an exposition of the present practice of the following firms: Messrs. J. Penn & Sons; Messrs. Maudslay, Sons & Field; Messrs. James Watt & Co.; Messrs. J. & G. Rennie. Messrs. R. Napier & Sons; Messrs J. & W. Dudgeon; Messrs. Ravenhill & Hodgson; Messrs Humphreys & Tenant; Mr J. T. Spencer, and Messrs. Forrester & Co. By N P. Burgh. Engineer. One thick 4to vol., cloth. $25 00; half morocco, 30 00

BACON. A TREATISE ON THE RICHARD'S STEAM-ENGINE INDICATOR — with directions for its use. By Charles T. Porter. Revised, with notes and large additions as developed by American Practice; with an Appendix containing useful formulæ and rules for Engineers. By F. W. Bacon, M. E. Illustrated Second edition. 12mo. Cloth $1.00; morocco, ... 1 50

ISHERWOOD ENGINEERING PRECEDENTS FOR STEAM MACHINERY. By B. F. Isherwood, Chief Engineer, U. S Navy With illustrations. Two vols. in one. 8vo, cloth, 2 50

STILLMAN. THE STEAM ENGINE INDICATOR — and the Improved Manometer Steam and Vacuum Gauges — their utility and application. By Paul Stillman. New edition. 12mo, cloth, ... 1 00

D. VAN NOSTRAND'S PUBLICATIONS.

MacCORD. A PRACTICAL TREATISE ON THE SLIDE VALVE, BY ECCENTRICS—examining by methods the action of the Eccentric upon the Slide Valve, and explaining the practical processes of laying out the movements, adapting the valve for its various duties in the steam-engine. By C. W. Mac Cord, A. M., Professor of Mechanical Drawing, Stevens' Institute of Technology, Hoboken, N. J. Illustrated. 4to, cloth. $3 00

PORTER. A TREATISE ON THE RICHARDS' STEAM-ENGINE INDICATOR, and the Development and Application of Force in the Steam-Engine. By Charles T. Porter. Third edition, revised and enlarged. Illustrated. 8vo, cloth, . . . 3 50

McCULLOCH A TREATISE ON THE MECHANICAL THEORY OF HEAT, AND ITS APPLICATIONS TO THE STEAM-ENGINE. By Prof. R. S. McCulloch, of the Washington and Lee University, Lexington. Va. 8vo, cloth, ... 3 50

VAN BUREN. INVESTIGATIONS OF FORMULAS—for the Strength of the Iron parts of Steam Machinery. By J. D. Van Buren, Jr., C. E. Illustrated. 8vo, cloth, . 2 00

STUART. HOW TO BECOME A SUCCESSFUL ENGINEER. Being Hints to Youths intending to adopt the Profession. By Bernard Stuart, Engineer Sixth edition 18mo, boards, ... 50

SHIELDS. A TREATISE ON ENGINEERING CONSTRUCTION. Embracing Discussions of the Principles involved, and Descriptions of the Material employed in Tunneling, Bridging, Canal and Road Building, etc., etc. By J. E. Shields, C. E. 12mo. cloth, 1 50

D. VAN NOSTRAND'S PUBLICATIONS.

WEYRAUCH. STRENGTH AND CALCULATION OF DIMENSIONS OF IRON AND STEEL CONSTRUCTIONS. Translated from the German of J. J. Weyrauch, Ph. D., with four folding Plates. 12mo, cloth, . . . $1 00

STUART. THE NAVAL DRY DOCKS OF THE UNITED STATES. By Charles B. Stuart, Engineer in Chief, U. S. Navy. Twenty-four engravings on steel. Fourth edition. 4to, cloth, ... 6 00

COLLINS. THE PRIVATE BOOK OF USEFUL ALLOYS, and Memoranda for Goldsmiths, Jewellers, etc. By James E. Collins. 18mo, flexible cloth, ... 50

TUNNER. A TREATISE ON ROLL-TURNING FOR THE MANUFACTURE OF IRON. By Peter Tunner. Translated by John B. Pearse. With numerous wood-cuts, 8vo, and a folio Atlas of 10 lithographed plates of Rolls, Measurements, &c. Cloth, . . 10 00

GRÜNER. THE MANUFACTURE OF STEEL. By M. L. Gruner. Translated from the French, by Lenox Smith, A.M., E.M.; with an Appendix on the Bessemer Process in the United States, by the translator. Illustrated by lithographed drawings and wood-cuts. 8vo, cloth, . . . 3 50

BARBA. THE USE OF STEEL IN CONSTRUCTION. Methods of Working, Applying, and Testing Plates and Bars. By J. Barba. Translated from the French, with a Preface by A. L. Holley, P.B. Illustrated. 12mo, cloth. 1 50

BELL. CHEMICAL PHENOMENA OF IRON SMELTING. An Experimental and Practical Examination of the Circumstances which Determine the Capacity of the Blast Furnace, the Temperature of the Air, and the Proper Condition of the Materials to be operated upon. By I. Lowthian Bell. 8vo, cloth, ... 6 00

D. VAN NOSTRAND'S PUBLICATIONS.

WARD. STEAM FOR THE MILLION. A Popular Treatise on Steam and its Application to the Useful Arts, especially to Navigation. By J. H. Ward, Commander U. S. Navy. 8vo, cloth, ... $1 00

CLARK. A MANUAL OF RULES, TABLES AND DATA FOR MECHANICAL ENGINEERS. Based on the most recent investigations. By Dan. Kinnear Clark. Illustrated with numerous diagrams. 1012 pages. 8vo. Cloth, $7 50; half morocco, ... 10 00

JOYNSON. THE METALS USED IN CONSTRUCTION: Iron, Steel, Bessemer Metals, etc, By F. H. Joynson. Illustrated. 12mo, cloth, ... 75

DODD. DICTIONARY OF MANUFACTURES, MINING, MACHINERY, AND THE INDUSTRIAL ARTS. By George Dodd. 12mo, cloth, 1 50

VON COTTA. TREATISE ON ORE DEPOSITS. By Bernhard Von Cotta, Freiburg, Saxony. Translated from the second German ed., by Frederick Prime, Jr., and revised by the author. With numerous illustrations. 8vo, cloth, ... 4 00

PLATTNER. MANUAL OF QUALITATIVE AND QUANTITATIVE ANALYSIS WITH THE BLOW-PIPE. From the last German edition. Revised and enlarged. By Prof. Th. Richter, o the Royal Saxon Mining Academy. Translated by Professor H. B. Cornwall. With eighty-seven wood-cuts and lithographic plate. Third edition, revised. 568 pp. 8vo, cloth, ... 5 00

PLYMPTON. THE BLOW-PIPE: A Guide to its Use in the Determination of Salts and Minerals. Compiled from various sources, by George W. Plympton, C. E., A. M., Professor of Physical Science in the Polytechnic Institute, Brooklyn, N. Y. 12mo, cloth, 1 50

D. VAN NOSTRAND'S PUBLICATIONS.

JANNETTAZ. A GUIDE TO THE DETERMINATION OF ROCKS; being an Introduction to Lithology. By Edward Jannettaz, Docteur des Sciences. Translated from the French by G. W. Plympton, Professor of Physical Science at Brooklyn Polytechnic Institute. 12mo, cloth, ... $1 50

MOTT. A PRACTICAL TREATISE ON CHEMISTRY (Qualitative and Quantitative Analysis), Stoichiometry, Blowpipe Analysis, Mineralogy, Assaying, Pharmaceutical Preparations. Human Secretions, Specific Gravities, Weights and Measures, etc., etc., etc. By Henry A. Mott, Jr., E. M., Ph. D. 650 pp. 8vo, cloth, ... 6 00

PYNCHON. INTRODUCTION TO CHEMICAL PHYSICS; Designed for the Use of Academies, Colleges, and High Schools. Illustrated with numerous engravings, and containing copious experiments, with directions for preparing them. By Thomas Ruggles Pynchon, D. D., M. A., President of Trinity College, Hartford. New edition, revised and enlarged. Crown 8vo, cloth, . . . 3 00

PRESCOTT. CHEMICAL EXAMINATION OF ALCOHOLIC LIQUORS. A Manual of the Constituents of the Distilled Spirits and Fermented Liquors of Commerce, and their Qualitative and Quantitative Determinations. By Alb. B. Prescott, Prof. of Chemistry, University of Michigan. 12mo, cloth, . 1 50

ELIOT AND STORER. A COMPENDIOUS MANUAL OF QUALITATIVE CHEMICAL ANALYSIS. By Charles W. Eliot and Frank H. Storer. Revised, with the co-operation of the Authors, by William Ripley Nichols, Professor of Chemistry in the Massachusetts Institute of Technology. New edition, revised. Illustrated. 12mo, cloth, ... 1 50

8

D. VAN NOSTRAND'S PUBLICATIONS.

NAQUET. LEGAL CHEMISTRY. A Guide to the Detection of Poisons, Falsification of Writings, Adulteration of Alimentary and Pharmaceutical Substances; Analysis of Ashes, and Examination of Hair, Coins, Fire-arms and Stains, as Applied to Chemical Jurisprudence. For the Use of Chemists, Physicians, Lawyers, Pharmacists, and Experts. Translated, with additions, including a List of Books and Memoirs on Toxicology, etc., from the French of A. Naquet, by J. P. Battershall, Ph. D.; with a Preface by C. F. Chandler, Ph. D., M. D., LL. D. Illustrated. 12mo, cloth, ... $2 00

PRESCOTT. OUTLINES OF PROXIMATE ORGANIC ANALYSIS for the Identification, Separation, and Quantitative Determination of the more commonly occurring Organic Compounds. By Albert B. Prescott, Professor of Chemistry, University of Michigan. 12mo, cloth, ... 1 75

DOUGLAS AND PRESCOTT. QUALITATIVE CHEMICAL ANALYSIS. A Guide in the Practical Study of Chemistry, and in the work of Analysis. By S. H. Douglas and A. B. Prescott; Professors in the University of Michigan. Second edition, revised. 8vo, cloth, ... 3 50

RAMMELSBERG. GUIDE TO A COURSE OF QUANTITATIVE CHEMICAL ANALYSIS, ESPECIALLY OF MINERALS AND FURNACE PRODUCTS. Illustrated by Examples. By C. F. Rammelsberg. Translated by J. Towler, M. D. 8vo, cloth, ... 2 25

BEILSTEIN. AN INTRODUCTION TO QUALITATIVE CHEMICAL ANALYSIS. By F. Beilstein. Third edition. Translated by I. J. Osbun. 12mo. cloth, ... 75

POPE. A Hand-book for Electricians and Operators. By Frank L. Pope. Ninth edition. Revised and enlarged, and fully illustrated. 8vo, cloth, ... 2 00

D. VAN NOSTRAND'S PUBLICATIONS.

SABINE. History and Progress of the Electric Telegraph, with Descriptions of some of the Apparatus. By Robert Sabine, C. E. Second edition. 12mo, cloth, . . $1 25

DAVIS AND RAE. Hand Book of Electrical Diagrams and Connections. By Charles H. Davis and Frank B. Rae. Illustrated with 32 full-page illustrations. Second edition. Oblong 8vo, cloth extra, . . . 2 00

HASKINS. The Galvanometer, and its Uses. A Manual for Electricians and Students. By C. H. Haskins. Illustrated. Pocket form, morocco, ... 1 50

LARRABEE. Cipher and Secret Letter and Telegraphic Code, with Hogg's Improvements. By C. S. Larrabee. 18mo, flexible cloth, ... 1 00

GILLMORE. Practical Treatise on Limes, Hydraulic Cement, and Mortars. By Q. A. Gillmore, Lt.-Col. U. S. Engineers, Brevet Major-General U. S. Army. Fifth edition, revised and enlarged. 8vo, cloth, 4 00

GILLMORE. Coignet Beton and other Artificial Stone. By Q. A. Gillmore, Lt. Col. U. S. Engineers, Brevet Major-General U. S. Army. Nine plates, views, etc. 8vo, cloth, ... 2 50

GILLMORE. A Practical Treatise on the Construction of Roads, Streets, and Pavements. By Q. A. Gillmore, Lt.-Col. U. S. Engineers, Brevet Major-General U. S. Army. Seventy illustrations. 12mo, clo., 2 00

GILLMORE. Report on Strength of the Building Stones in the United States, etc. 8vo, cloth, ... 1 00

HOLLEY. American and European Railway Practice, in the Economical Generation of Steam. By Alexander L. Holley. B. P. With 77 lithographed plates. Folio, cloth, 12 00

www.ingramcontent.com/pod-product-compliance
Lightning Source LLC
Chambersburg PA
CBHW020054170426
43199CB00009B/284